NEXUS NETWORK JOURNAL Architecture and Mathematics

Aims and Scope

Founded in 1999, the *Nexus Network Journal* (NNJ) is a peer-reviewed journal for researchers, professionals and students engaged in the study of the application of mathematical principles to architectural design. Its goal is to present the broadest possible consideration of all aspects of the relationships between architecture and mathematics, including landscape architecture and urban design.

Editorial Office

Editor-in-Chief
Kim Williams
Corso Regina Margherita, 72
10153 Turin (Torino), Italy
E-mail: kwb@kimwilliamsbooks.com

Contributing Editors
The Geometer's Angle
Rachel Fletcher
113 Division St.
Great Barrington, MA 01230, USA
E-mail: rfletch@bcn.net

Book Reviews
Sylvie Duvernoy
Via Benozzo Gozzoli, 26
50124 Firenze, Italy
E-mail: syld@kimwilliamsbooks.com

Corresponding Editors
Alessandra Capanna
Via della Bufalotta 67
00139 Roma Italy
E-mail: alessandra.capanna@uniroma1.it

Tomás García Salgado
Palacio de Versalles # 200
Col. Lomas Reforma, c.p. 11930
México D.F., Mexico
E-mail: tgsalgado@perspectivegeometry.com

Robert Kirkbride
studio 'patafisico
12 West 29 #2
New York, NY 10001, USA
E-mail: kirkbrir@newschool.edu

Andrew I-Kang Li
Initia Senju Akebonocho 1313
Senju Akebonocho 40-1
Adachi-ku
Tokyo 120-0023 Japan
E-mail: i@andrew.li

Michael J. Ostwald
School of Architecture and Built Environment
Faculty of Engineering and Built Environment
University of Newcastle
New South Wales, Australia 2308
E-mail: michael.ostwald@newcastle.edu.au

Cover
Nexus 2010 Logo
Gonçalo Azevedo

Vera Spinadel
The Mathematics & Design Association
José M. Paz 1131 - Florida (1602), Buenos Aires, Argentina
E-mail: vspinade@fibertel.com.ar

Igor Verner
The Department of Education in Technology and Science
Technion - Israel Institute of Technology
Haifa 32000, Israel
E-mail: ttrigor@techunix.technion.ac.il

Stephen R. Wassell
Department of Mathematical Sciences
Sweet Briar College, Sweet Briar, Virginia 24595, USA
E-mail: wassell@sbc.edu

João Pedro Xavier
Faculdade de Arquitectura da Universidade do Porto
Rua do Gólgota 215, 4150-755 Porto, Portugal
E-mail: jpx@arq.up.pt

Instructions for Authors

Authorship
Submission of a manuscript implies:
• that the work described has not been published before;
• that it is not under consideration for publication elsewhere;
• that its publication has been approved by all coauthors, if any, as well as by the responsible authorities at the institute where the work has been carried out;
• that, if and when the manuscript is accepted for publication, the authors agree to automatically transfer the copyright to the publisher; and
• that the manuscript will not be published elsewhere in any language without the consent of the copyright holder.
Exceptions of the above have to be discussed before the manuscript is processed. The manuscript should be written in English.

Submission of the Manuscript

Material should be sent to Kim Williams
via e-mail to: kwb@kimwilliamsbooks.com
or via regular mail to: Kim Williams Books,
Corso Regina Margherita, 72,
10153 Turin (Torino), Italy

Please include a cover sheet with name of author(s), title or profession (if applicable), physical address, e-mail address, abstract, and key word list.

Contributions will be accepted for consideration to the following sections in the journal: research articles, didactics, viewpoints, book reviews, conference and exhibits reports.

Final PDF files
Authors receive a pdf file of their contribution in its final form. Orders for additional printed reprints must be placed with the Publisher when returning the corrected proofs. Delayed reprint orders are treated as special orders, for which charges are appreciably higher. Reprints are not to be sold.

Articles will be freely accessible on our online platform SpringerLink two years after the year of publication.

Nexus Network Journal

QUALITATIVE AND QUANTITATIVE ARCHITECTURE AND MATHEMATICS

VOLUME 13, NUMBER 2

Summer 2011

KIM WILLIAMS BOOKS

Nexus Network Journal

QUALITATIVE AND QUANTITATIVE
ARCHITECTURAL RESEARCH IN MATHEMATICS

VOLUME 16, NUMBER 3
Winter 2014

Nexus Network Journal
Vol. 13
No. 2
Pp. 255–502
ISSN 1590-5896

CONTENTS

Kim Williams | *Letter from the Editor*

Kim Williams Books
Corso Regina Margherita, 27
10153 Turin (Torino) ITALY
kwb@kimwilliamsbooks.com

Abstract. *NNJ* Editor-in-Chief introduces *Nexus Network Journal* volume 13 number 2 (Summer 2011).

Keywords: Nexus Network Journal, Nexus conferences

Following the seventh edition of the conference series "Nexus: Relationships between Architecture and Mathematics" (San Diego, 2008), it was decided that papers presented at future editions of the conference would no longer be published in separate volumes as conference proceedings, but would be appear instead in the *Nexus Network Journal.* Thus the papers presented at the eighth edition of the Nexus conference (Porto, 2010) are being published in the three issues of this present vol. 13. The first group appeared in the first issue (Winter 2011), dedicated to "Shape and Shape Grammars", under the direction of Lionel March. This present issue, Summer 2011, contains papers presented during two sessions moderated by myself and Sylvie Duvernoy. The final issue of this year, Autumn 2011, will contain papers presented during two sessions directed by José Calvo Lopez (From Mediaeval Stonecutting to Projective Geometry) and Gonçalo Furtado (Architecture, Systems Research and Computational Sciences).

The two sessions included here comprised presentations dedicated to a wide range of subjects rather than one specific topic, and represent the kind of variety that has always characterized Nexus conferences, while at the same time making evident the many ways in which what seem to be very different topics are intertwined. I have entitled this issue "Qualitative and Quantitative Architecture and Mathematics." And what a rich variety it is! The issue opens with papers by two of the three keynote speakers. Lino Cabezas provides an in-depth examination of "Ornamentation and Structure in the Representation of Renaissance Architecture in Spain", and Chris Williams discusses "Patterns on a Surface: The Reconciliation of the Circle and the Square." At first glance, these two themes could not appear more antithetical, the one dealing with perspective analyses of decorated surfaces in paintings, and the other with a numeric method for tiling a curved surface by relaxing a grid over it. However, in his own way each author is dealing with the application of geometrical shapes on architectural surfaces. (The third keynote speaker at Nexus 2010 was Eduardo Souto de Moura, winner of the 2011 edition of the Pritzker Prize, whose works speak for themselves regarding the connection between architecture and mathematics.)

Three of the papers included in this issue concern housing. Marco Giorgio Bevilacqua presents "Alexander Klein and the *Existenzminimum*: A 'Scientific' Approach to Design Techniques", showing how rigorous mathematical methods of analysis were used to develop standards for decent housing for the working class. In "The Octagon in the Houses of Orson Fowler", Eliseu Gonçalves looks at how the same search for decent, comfortable and affordable housing led American phrenologist Orson Fowler to champion the octagon as the ideal shape for a house. This ties into the theme of the Geometer's Angle column in this issue, "Thomas Jefferson's Poplar Forest", where with her usual care geometer Rachel Fletcher uses the original plan for Jefferson's octagonal house to highlight the use of root-two proportions in its design.

Moving from the domestic scale to that of the city, two of the papers here deal with urban design. In "Measuring Lisbon Patterns: Baixa from 1650 to 2010," Teresa Marat-

DOI 10.1007/s00004-011-0066-4; *published online* 8 June 2011

Mendes, Mafalda Teixeira de Sampayo and David M.S. Rodrigues present the results of a mathematical-urban analysis of the evolution of Lisbon's historic Baixa district over the course of 350 years. In "Dynamics of Urban Centre and Concepts of Symmetry: Centroid and Weighted Mean", Jong-Jin Park introduces the concept of the programmatic moving centre in order to use the mathematical methods of analysis of centroid and weighted mean to describe the generation of urban form.

It hardly bears repeating that scientific thought in Renaissance and Baroque periods had an enormous influence on architectural theory. Documentary evidence of this is found in both written treatises and built work. Three papers in this issue are concerned with the sixteenth, seventeenth and very early eighteenth centuries. In "From Quantitative to Qualitative Architecture in the Sixteenth and Seventeenth Centuries: A New Musical Perspective", Vasco Zara examines the 1499 *Hypnerotomachia Poliphili* and the 1679 *Architecture Harmonique* in support of a new interpretation of the use of proportion and a transition between two differents concepts of architecture that took place in the sixteenth century. In "Renaissance in Goa: Proportional Systems in Two Churches of the Sixteenth Century" António Nunes Pereira presents the results of geometric and proportional analyses of two churches to demonstrate the use of Renaissance proportional systems in Portuguese India. But the new science of perspective that was developed beginning in the fifteenth century brought with it an awareness of the contrast between true proportion and visual perception. João Paulo Cabeleira Marques Coelho, in "Inácio Vieira: Optics and Perspective as Instruments towards a Sensitive Space", discusses two early eighteenth-century treatises by the Portuguese Jesuit scholar in which he searches for *prós opsin euruthmia* (proportion in agreement with visual impression).

Completing this issue are three fascinating research papers and a book review. In "Making a Difference," archaeologist Lone Mogensen proposes a theory about how the ground plans of medieval churches were initially staked out. Michael Ostwald presents the first of a two-part study (the second part to appear in the next issue of the *NNJ*) entitled "The Mathematics of Spatial Configuration: Revisiting, Revising and Critiquing Justified Plan Graph Theory" in which he provides a history, explanation and worked example of analysis using JPG. As he did in his earlier study of applications of fractal theory to architecture (*NNJ* 3, 1 (2001): 73-84), one of the most widely read articles ever published in our journal, Michael steps back and looks at the subject to present an overview that is both dispassionate and rigorous. Bernard Parzysz points out interesting solutions for transitions between geometric shapes in "From One Polygon to Another: A Distinctive Feature of Some Ottoman Minarets". Finally, Book Review Editor Sylvie Duvernoy reviews *The Mathematical Works of Leon Battista Alberti*, a book about which I happen to know a great deal, since it is the result of a two-year effort on the part of Stephen Wassell, Lionel March and myself to compile translations and commentaries of four mathematical works by the great Renaissance polymath.

As always, it is an honor and privilege to present such a fine group of papers. Happy reading,

Kim Williams

Lino Cabezas

Departamento de Dibujo
Universidad de Barcelona
Diagonal Sud
Facultat de Belles Arts
Pau Gargallo, 4
08028 Barcelona SPAIN
linocabezas@ub.edu

Research

Ornamentation and Structure in the Representation of Renaissance Architecture in Spain

Presented at Nexus 2010: Relationships Between Architecture and Mathematics, Porto, 13-15 June 2010.

Keywords: perspective theory, Pedro Berruguete, Juan Bautista Villalpando, vaults, coffered ceilings, Renaissance architecture, ornamentation

Abstract. In the Renaissance, geometric perspectives became a method of architectural reasoning. We look at two works: a painting attributed to Pedro Berruguete and an engraving by Juan Bautista Villalpando in order to to discover the complex relationship between geometry and architecture, and in particular, the form of representation. Our analysis is based on the idea that geometric perspective meant architecture of the period was capable of graphically controlling reality through a high degree of geometric precision.

Images are often the most important references for the discovery of relations between geometry and architecture. Some images may attain extraordinary prominence and acquire total autonomy, even becoming the centre of attention in aesthetic debates and when formulating theories in general. Using images from different historical situations, it is possible to discover many aspects related to architecture.

In the case of images of architecture from the beginning of the Modern Era, this is easy to understand: at that time, the culture of *Disegno* was at its peak in the history of art, in parallel with the construction of a science of representation and the appearance of a new art form driven by artists' thirst for invention.

In this Renaissance setting, the formulation of geometric perspectives was of major importance, and its use in architecture caused a revolution with clear results, even though they have not always been considered good. The architect Bruno Zevi, recognising its significance, gave a very negative evaluation of the consequences of this visual culture of perspective. In *Il linguaggio moderno dell' architettura*, he wrote:

> In the early fifteenth century disaster struck. It was the triumph of perspective. Architects stopped worrying about architecture, and just drew it. The damage was enormous, increased over the centuries and continues to grow with industrialised construction [Zevi 1973; our trans.].

While the words of Zevi are conditioned by the historic circumstances of his time, they serve to illustrate a debate in which favourable opinions on perspective are more numerous. For many, since the Renaissance, in addition to being the expression of a powerful visual conception of the arts, perspective has become a method for architectural reasoning. This is like the *species dispositionis* concept of the Roman architect Vitruvius, who, in the first century B.C., defined perspective (*scenographia*), together with the floor plan (*ichnographia*) and elevation (*ortographia*), as three kinds of architectural arrangement, and not just a simple graphic system to represent it.

Before focussing on the analysis of architecture represented in perspective, we are aware of the difficulty in indiscriminately applying the term 'image' to all representations of architecture. According to René Huyghe [1965], our culture has evolved from the 'Civilisation of the book' into the 'Civilisation of the image'. For this reason, we highlight our interest in one of the special functions of images in architecture, that which is conditioned or directed towards its graphic control, to resolve the geometry.

Geographically and chronologically, we focus on architecture in Spain around the sixteenth century. We look at two works: a painting attributed to Pedro Berruguete (fig. 1), dating back to the end of the fifteenth century, and an engraving by Juan Bautista Villalpando (fig. 2), from the end of the sixteenth century. Our interest in both is to discover the complexity of the relationship between geometry and architecture, and in particular, the form of representation.

In the hundred years which separate the two works, there were major changes not only in architecture and the theoretical development of geometry of form, but also major transformations in design approaches and the construction techniques used during this, the sixteenth, century.

In the painting we focus our attention on the Mudejar coffered ceiling of the chapel, while in the engraving we are interested in the groin vault, decorated with rhomboidal coffering. The images have in common a visual conception of architecture, demonstrated by the geometric perspective which would mature scientifically in Europe during the sixteenth century, with contributions from important writers such as Viator [1505], Dürer [1525], Serlio [1545] and Vignola [1583] (fig. 3).

In the panel of Pedro Berruguete, different formal and stylistic elements of the time can be identified. The architectonics of the composition has a perspective which gives a natural feeling to a chapel which takes centre stage for its appearance of reality. The confluence of traditions in this is clear, at a time which contemporary history describes a Spain in transition from Hispano-Flemish art to painting in line with the postulates of Italian art.

The similarities of the Gothic fretwork to a panel by Rogier van der Weyden are evident, continuing with tradition (fig. 4). The Gothic elements contrast stylistically with the lateral semi-circular arches and the pseudo-Corinthian capitals which support them. It was painted at a time when Gothic forms coexisted with the new Renaissance models; under these circumstances, the ability to integrate different styles was considered a professional merit which many artists could boast of. Despite this, the new 'Roman' Renaissance repertoires, written about in detail, circulated and were imposed, demanding a purity of style incompatible with previous forms. In 1526, in the first treatise on art printed in Spain, its author, Diego de Sagredo, condemns those conceited enough to boast of their ability to mix the new Renaissance repertoires with the Gothic tradition which was still alive and which he wanted to overthrow: "And take care not to presume to mix Roman with modern [i.e., Gothic]" [Sagredo 1526: iiii r].

Leaving aside the clash between Gothic and Renaissance, latent in the panel, the presence of a Mudejar ceiling is the most remarkable feature of the painting. This element is of a firmly seated typology of the time, in line with the abundance of these wooden structures in architecture in most of the Iberian peninsula.

Our analysis is based on the idea that geometric perspective meant architecture of the period was capable of graphically controlling reality through a high degree of geometric precision. However, examining the graphic precision of the painting attributed to Berruguete, there is a contradiction between the low level of geometric precision in the spatial construction of the chapel and the higher level in the work on the coffered ceiling (fig. 5).

Fig. 1. Pedro Berruguete, *Virgen con el niño*. Madrid. Museo Municipal

Fig. 2 Juan Bautista Villalpando, perspective of the inside of the Holy of Holies, *In Ezechielem Explanaciones*, 1595

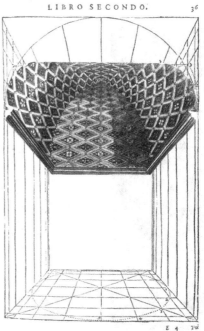

Fig. 3. Photomontage of the works analysed and the engraving from Sebastiano Serlio, *Libro de perspectiva*, 1545

Fig. 4. Rogier Van der Weyden, La *Virgen con el Niño entronizada.*
Museo Thyssen- Bornemisza, Madrid

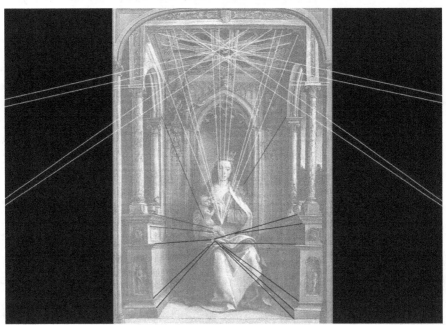

Fig. 5. Geometric analysis of the perspective in the panel by Pedro Berruguete

The painting is a small panel kept in the Madrid Municipal Museum. The importance of Pedro Berruguete is rooted in having been identified by art critics and historians as marking the transition to a period of great art in Spain that would reach its peak in the seventeenth century, the Golden Age of Spanish painting. His role in introducing a new Italianised style, far removed from Flemish models, was made possible, according to certain critics, by a supposed stay in Urbino, working for the duke and in contact with Piero della Francesca and Francesco di Giorgio Martini.[1]

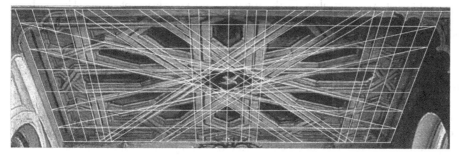

Fig. 6. Analysis of the geometry of the coffered ceiling of the chapel in the panel by Pedro Berruguete

However, geometric analyses of other paintings, without doubt by Berruguete, do not have the characteristics or the high quality of Italian perspective present in works created in Piero's circles. In the panel of the Spanish painter, the inconsistencies in perspective evident in setting up the main volume contrast with a much higher level of precision in the geometric tracing of the Mudejar coffered ceiling (fig. 6). For this reason we feel that the quality of the representation of the ceiling should be attributed to his knowledge of this building technique itself, rather than the hypothetical mastery of specific methods of perspective of Italian art at the end of the fifteenth century. In our interpretation, we take into account the major conceptual difference between the geometry used in construction and that for perspective representation, developed in Italy for visual control of pictorial space.

Fig. 7. Design of interlocking lines laid out in a pattern of eight, in different materials and eras

It is well known that Spanish Mudejar coffered ceilings achieved a technical maturity in the sixteenth century due to the rigorous organisational structure of the professionals. Their geometric techniques, something peculiarly their own, were based on highly sophisticated plans, preserved today in manuscripts and writings. While the wooden structures of the coffered ceiling were considered to be of major quantitative and qualitative importance among these craftsmen, we know that the same geometry was used in other professions and in other materials, such as plaster, stone and ceramics (fig. 7), media that were not alien to the interests and knowledge of artists. Under these circumstances, it is not surprising that paintings such as the panel by Berruguete exist, which include Mudejar ornamentation.

The experience of geometric techniques applied to fretwork had been accumulated over centuries in the heart of the trade guilds, very different from the situation which framed the development of geometric perspective applied by artists early in the Renaissance era, with characteristics much more 'intellectual' in their theoretical formulation. In the case of Berruguete's panel there is a singular phenomenon: the perspective of the coffered ceiling is foreshortened, which means that the method used for painting it was probably not identical to that used by craftsmen to draw up plans for the construction of wooden coffered ceilings.

While the method of drawing is unknown, it seems that the perspective of the coffered ceiling could have been drawn without necessarily using a vanishing point for the main volume: it is possible to limit the geometric tracing of the coffered ceiling to the area occupied in the painting. We tested a hypothesis for drawing the perspective without using a vanishing point, based on a modular structure (fig. 8).

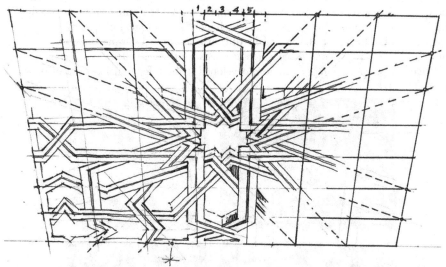

Fig. 8. Hypothesis for drawing the perspective of a coffered ceiling using a modular grill

The preparatory drawings Berruguete is known to have made on some panels, by carving grooves onto the surface to mark the architectural lines, confirm the hypothesis that, on many occasions, the geometry used by artists is basically ornamental, as a filler, and not a spatial or design geometry. Despite this, the final effect is of high quality, demonstrating the painters' capability to synthesise and harmonise very heterogeneous elements to achieve a unitary figurative conception.

There are antecedents to Gothic panels where the ornamental geometry for filling the background is solved in the same way as in traditional constructions of coffered ceilings, without applying distortions to enhance perspective (fig. 9). Although not much evidence is conserved of the geometric methods used by the professionals, the documents available demonstrate a greater stability and autonomy of craftsmanship, compared to the complexity and constant renovation of methods of perspective.

Fig. 9. Gothic panel by unknown author, and Mudejar coffered ceiling with identical interlocking design

Leaving aside the circumstances, the transformations produced during the sixteenth century were major and led to the innovation of formulas to adapt the traditional repertoires of the trades to a new visual conception of architecture and of the arts in general. We know that the highly skilled stonemasons knew how to foreshorten structures such as coffered ceilings. One of the most interesting documents from the century relating to this problem is the *Libro de arquitectura,* or *Book of Architecture,* by Hernán Ruiz the Younger, who resolved the challenge of adapting a regular surface of coffers to a trapezoid splayed lintel (fig. 10). On this subject, it should be recognised that, in Spain, development of the geometry was under scientific dominance and the authority of architect-designers.

Thanks to the development of printing, architecture not only attained great prestige due to the actual buildings, but also as a result of the high quality and increasingly precise printed images. An example of this is found in a letter from de Herrera to Cristóbal de Salazar, written when the architect was about to complete works on the Escorial Palace Royal Monastery: "It seemed right to show the whole world a brickwork structure of such great significance" (cited in [Llaguno y Amirola, 1829; our trans.]).

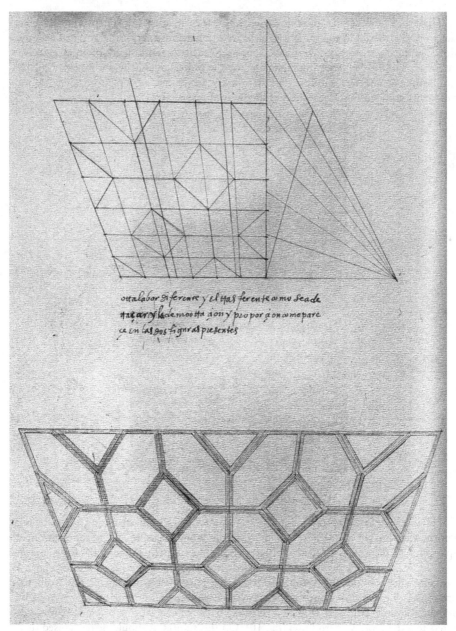

Fig. 10. Hernán Ruiz the Younger, *Libro de arquitectura,* Treatise on how to draw, present plans and establish proportion

He was referring to the imminent publication of the *Sumario y breve declaración de los diseños y estampas de la fábrica de San Lorenzo el Real del Escorial* (Summary and brief declaration of the designs and illustrations of the building of San Lorenzo el Real del Escorial) [Herrera 1589], which was illustrated with a famous perspective, or '*scenographia*' (fig. 11).

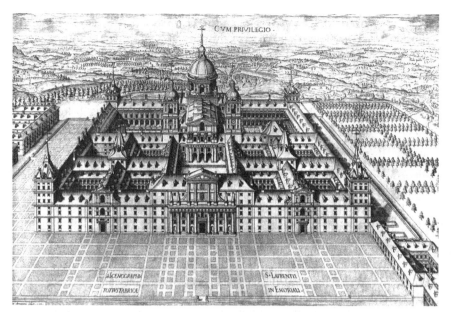

Fig. 11. Juan de Herrera, "Séptimo sello", "Scenographia" of the Escorial Palace Monastery, from *Sumario y breve declaración de los diseños...* [1589]

However, images of architecture were not limited to representing or demonstrating works already carried out. They also took on the function, increasingly important over time, of projecting new constructions. In both cases, the requirement for graphic rigour and conceptual precision of geometric forms is the reason for the advance in many theories of geometry of form. Little by little, the images which combined tradition and innovation, experience and scientific discovery became the stimulus and starting point for graphic considerations in the new architecture.

The other image we analysed, the engraving by Villalpando, accompanied the most extensive and erudite theoretical discourse in Spain on the relationship between perspective and architecture. This discourse developed specifically in the context of the analysis of Vitruvian graphics for the reconstruction of the *Temple of Solomon* project recorded by the prophet Ezekiel. Juan Bautista Villalpando refers to this in *Ezechielem explanationes et apparatus urbis ac templi Hierosolymitani* [Villalpando and Pradi 1596-1604].

In this text, what draws attention is the disproportion between the theoretical arguments used, mainly philological and geometric, as well as theological, in contrast to the few arguments based on professional architecture in relation to drawing plans. A fragment of the text demonstrates the difficulties of creating a perspective of the temple due to the high level of scientific requirements, impossible in the everyday work of an architect.

> ...the perspective demands so much diligence, so much art and work, enough to intimidate many, even the wisest and most learned, so that they give up. Well I, who have spent much of my time carrying out this work, without giving in to fatigue, sparing no costs or effort, from the very beginning, have been dominated by the unique wish to show, among so many graphic reconstructions of the Temple, a perspective of the whole building. I have tried

to do it many times, but have given up even more often because of the difficulty, the time required, defeated by the delay in completing it. However, I who have not been able to present the perspective of the whole Temple, present here that of its major and most sacred section, namely that of the Holy of Holies, with this inscription: PERSPECTIVE OF THE VAULT OF THE WALLS AND THE PAVING OF THE SANCTA SANTORUM AND OF THE ARC OF THE COVENANT WITH THE CHERUBIM [Villalpando 1990: 166].

The demands Villalpando imposed on his own graphic work become evident when he confesses having tried to draw a perspective of the whole Temple, without being able to finish it, settling for the perspective of the "major and most sacred section": the Holy of Holies, on which we focus our attention. He dedicated more than fifteen years of work before he published this perspective of the whole Temple, never completed. However, after confessing that he had tried but stopped many times because of the difficulties involved, Villalpando adds: "I am not without hope that I will be able to offer the perspective of the whole Temple after giving the final touch to this work, if God gives me the strength and long life" [Villalpando 1990: 166].

There were two challenges for drawing the perspective of the Holy of Holies. First there was the groin vault, formed by the intersection of two cylindrical surfaces, defining two ellipses, whose graphic resolution using a series of points had been perfectly worked out by Piero della Francesca more than a century earlier [Piero della Francesca 1984: 122]. And even though Piero's *De prospectiva pingendi* was never published, the scheme became well-known through the treatises of others, such as that of Sebastiano Serlio [1545].

The second geometric problem of Villalpando's engraving consists in the construction of the rhomboids on the paving, walls and vault. Nowadays, in contrast to what was known in the sixteenth century, we note geometric 'errors' in the engraving of the Holy of Holies which are easily detectable with present-day knowledge that allows us to distinguish, on the surface, straight generatrices, ellipses and cylindrical helices (fig. 12). All these questions of representation and graphic control of architecture were then resolved intuitively, as in the case of the treatise by Serlio.

When it comes to the 'style' of the rhomboidal coffers which appear in Villalpando's engraving, it is obviously indebted to the monuments of Roman architecture, particularly in one of the niches conserved in the apse of the second-century Temple of Venus and Rome (fig. 13). The engraving was made in Rome around 1594-1595, four years after Villalpando arrived in the city. Roman architecture was a major influence throughout Europe at that time, and specific elements, such as the Roman niche, became well-known thanks to their influence on other constructions and their impact on illustrations in writings such as those of Philibert de l'Orme [1567] and Androuet du Cerceau [1576-1579]. In these publications, the authors reproduced the Royal Chapel of the Château d'Anet (figs. 14 and 15), the most famous work by de l'Orme, with rhomboid coffers, done after his three-year stay in Rome, where he had, according to Blunt, studied, measured, and even excavated in the Roman ruins [Blunt 1958: 14].

At the time, with the geometric method of design, a complex spatial structure could be represented in two dimensions, such as the vault in the Royal Chapel of the Château d'Anet where the form of the paving is similar to that of the vault, and they can be associated using straight lines perpendicular to the floor. In one of his treatises (fig. 16), de l'Orme acknowledges how proud he is of the geometric method he applied to the project:

Fig. 12. Theoretical analysis of the lines in the Vilalpando engraving. A: ellipsis, B: straight line, C: cylindrical helix

Fig. 13. Apse of the Temple of Venus and Roma, s. II, Rome

Fig. 14. Philibert de l'Orme, *Royal Chapel of Château de Anet*, 1567

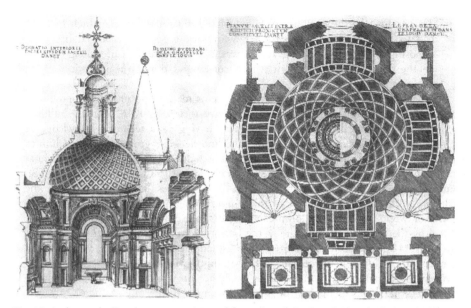

Fig. 15. Androuet du Cerceau, cross section and floor plan of the *Royal Chapel of Château de* Anet [1576-1579]

Fig. 16. Comparison of the paving and the vaulted ceiling of the *Château de Anet* Royal Chapel, designed by Philibert de l'Orme

Those who take the trouble will know what I did for the dome which I had to build in the chapel of Anet, with various sections arranged in opposing directions. The same system was used to create the coffer design on the floor and the pavement, which correspond through a projection along perpendicular lines.[2]

The maturity and capacity of the graphic systems made it possible to propose new and complex forms which did not necessarily have their origins in practical building experience. In the case of Dürer, in addition to having solid technical knowledge of architecture, his mastery of graphic resources and knowledge of geometry allowed him to generate new forms which could not only be used in architecture, but were also valued as aesthetically pleasing forms (fig. 17). The works of Plato affirm that, in Ancient Greece, geometric shapes were considered beautiful in themselves, as they related to the intelligible world of ideas. This way of thinking explains the existence, in the sixteenth century, of works such as that of the mannerist goldsmith Wentzel Jamnitzer who, obsessed by the symbolic architecture of the universe, created geometric figures of great beauty (fig. 18). The consequence of this aesthetic tradition was to materialise in Baroque, in the work of architects such as Borromini (fig. 19).

Starting in Renaissance, printed images would acquire the status of formal proposals and become collections of models for use by craftsmen. In the case of Viator's treatise, the first publication dedicated to perspective, the author describes the figures as *exemples ou inductives de besonger en plus grandes* (examples or encouragement to be carried out on a larger scale) [Viator 1505] (fig. 20). However, using an architectural drawing as the base for a painting, another flat image, was different to transferring it to stone or other construction material. In the sixteenth century, the efficiency of stonework was based on technical perfection of the structures, where the classification of elements such as springers, groins and infilling was governed by their mechanical efficiency, with ornamentation playing a complementary role.

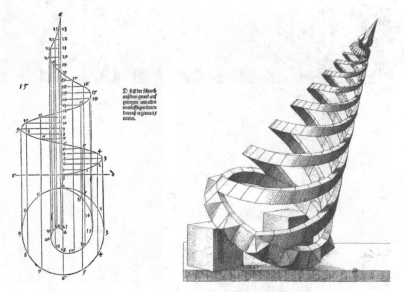

Fig. 17 (left). Albrecht Dürer, conical helix. From his treatise on geometry,
Underweysung der Messung, 1525
Fig. 18 (right). Wentzel Jamnitzer, helicoid. In *Perspectiva Corporum Regularium*, 1568

Fig. 19. Borromini, sketches of Sant'Ivo alla Sapienza, 1642-1660

Fig. 20. Viator, Engraving in *De artificiali perspectiva*, 1505

On some occasions, the new ornamental repertoires, such as the coffer designs taken from Roman architecture, were adapted to the construction of the masonry structure. A particular example is the vault of the passage to the sacristy in the cathedral in Murcia, with movement generated by a semi-circular structure supported by curved walls. In this vault, each wedge-shaped voussoir of the stone structure coincides with two sunken panels, with very precise contact between structure and ornament, giving the appearance of a coffered surface made up of different pieces, as are the voussoirs (fig. 21).

Fig. 21. Coffered vault in the passage of the vestry in the cathedral in Murcia

Contrasting with this unity of structure and ornament, a very obvious separation is present in other constructions. The first theoretical work by Philibert de l'Orme, *Le premier Tome de l'Architecture* [1561], is a clear example of a treatise dedicated to exclusively structural elements. This was the first French publication on architecture dedicated to construction, particularly to the light wooden structures he invented himself, which did not require large joists (fig. 22). The ornamentation was independent and left to a later phase of construction.

The wooden vaults proposed by de l'Orme were built exactly as he had designed them, both in Europe and abroad. A well-known example is the wooden vaulted ceiling in the Jesuit church in the Argentinean city of Rosario (fig. 23). The visible sides of the wooden elements were decorated in various ornamental styles, often the responsibility of craftsmen other than those involved in the construction.

De l'Orme's interest changed from the purely structural in his first treatise to matters of style and ornamentation in his second, the much more theoretical *Premier tome de l'architecture* [1567] published six years later. In this, his focus on style led him to add a sixth architectural rule, the 'French' rule, to the five existing principles. This solved a technical problem which prevented monolithic blocks being extracted from French quarries for column shafts and caused them to be built them using smaller blocks.

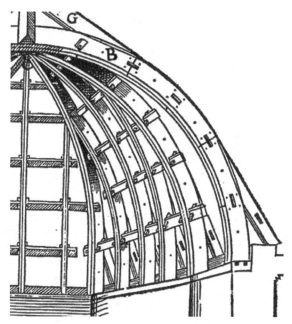

Fig. 22. Wood structure of a vaulted ceiling
from *Nouvelles inventions pour bien bastir et a petits fraiz,* by Philibert de l'Orme

Fig. 23. The upper convex surface of the wooden vaulted ceiling of the Jesuit church
in the city of Rosario (Argentina)

The architect devised tambours with a ring design on the surface to hide the joints in the shaft, thus giving priority to the new style over structural 'truth'. In the following century, giving precedence to innovation over masonry construction techniques was to become more widespread; the Spaniard Juan Andrés Ricci proposed using ornamentation to expressly hide the joints between stones, considering this one of the most important questions in architecture:

> One of the most beautiful aspects of architecture is hiding the joints, not only through ashlar masonry where they are set back, but also in the bases and capitals, and with mitre joints, so that joints will not be seen as obviously as they are today, in cornices and other important areas, an appalling thing, meriting repair (cited in [Berchez and Marías 2002, our trans.].

Returning to the case of Villalpando's vaults, we see that, in one of his engravings he uses light wooden structures, in the representation of the section of the Santuario where the suspended vault is hung from the structure, a solution justifiable only for its symbolic value or for the purely ornamental function of its elements (fig. 24).

Fig. 24. Juan Bautista Villalpando, section of the Sanctuary with the suspended vault ceiling, 1595

The system of construction was similar to those proposed by de l'Orme in his treatise, wooden vaults which would become generally used in Spain in the seventeenth century. Fray Laurencio de San Nicolás expressly commented on this phenomenon in his excellent treatise on architecture:

In Spain, particularly in this Court, covering chapels with wooden domes is being introduced, and it is a very safe and strong structure, looking like stone on the outside" [Laurencio of Saint Nicholas 1664: ch. LI, 189; our trans.].

The Spanish name *bóvedas encamonadas* [false vaults] is used for those made in lightweight materials such as wood or reed, covered with plaster, usually so that ornamentation can be painted on. This technique had a mediaeval background, and a type of covering with structures similar to those of Philibert de l'Orme was being used in Spain. This was based on Mudejar carpentry where it was normal to use a technique called '*ataurejadas*', in which the ornamentation was not defined by the elements of the structure itself but as a result of attaching interlaced strips of wood on the wooden panels covering the framework to obtain an interlocking, geometric ornamentation (fig. 25).

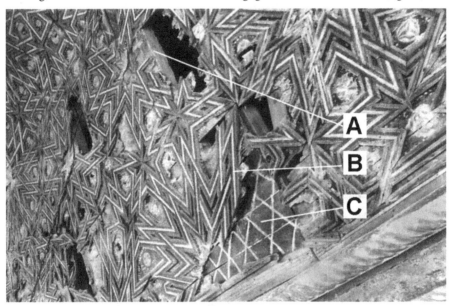

Fig. 25. Detail of the coffered ceiling of Santa Colomba de la Vega (León): A, "pares" (small sloping joists); B, "taujeles" (interlaced carpentry); C, "tablazón" (panelling)

In the professional and stylistic situation of the sixteenth century, the images of Villalpando's work were used, in a theological interpretation, as an instrument to create emotions, or, as the expression of the scientific instrument of *Disegno* for construction projects. In the vault of the etching of the *Sancta Sanctorum* it is clear that the ribbing has no structural function but is purely ornamental.

With the invention of perspective geometry, and the science of *Disegno* in general, the function of images was to design pure forms, often independent of the material factors. This can be seen in a comment by Leonardo da Vinci on a famous drawing in the *Codex Atlanticus*: "Corpo nato della prospective" (fig. 26). This graphic proposal by Leonardo would be taken up and made their own by later authors, although without it being used directly in construction. Almost eighty years later, we recognise a similar figure in a treatise on perspective by Daniele Barbaro, published in 1569 (fig. 27), and four centuries later another by the artist Escher in one of his famous engravings (fig. 28).

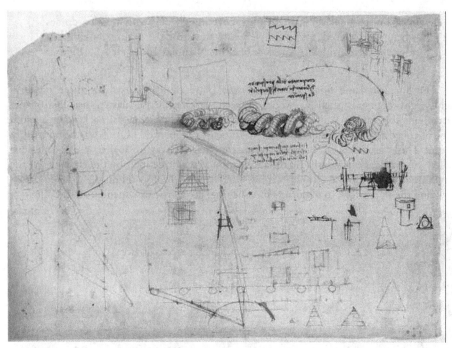

Fig. 26. Leonardo da Vinci, *Códice Atlántico*, c. 1490: "Corpo nato della Prospettiva"

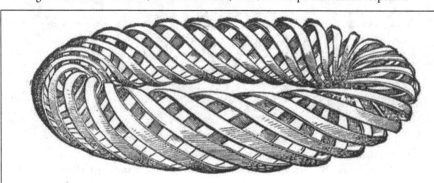

PARTE SECONDA
Nellaquale fi tratta della Ichnographia ,
cioè defcrittione delle piante.

PRATICA DI DESCRIVERE LE FIGVRE
di molti anguli in uno circolo.   Cap. I.

Fig. 27. Daniele Barbaro, *La pratica della Perspectiva*, 1569

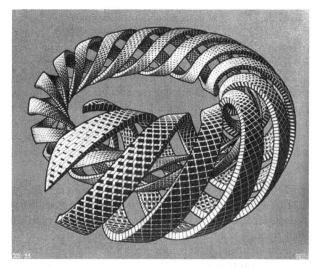

Fig. 28. M.C. Escher, *Spirals*, 1953

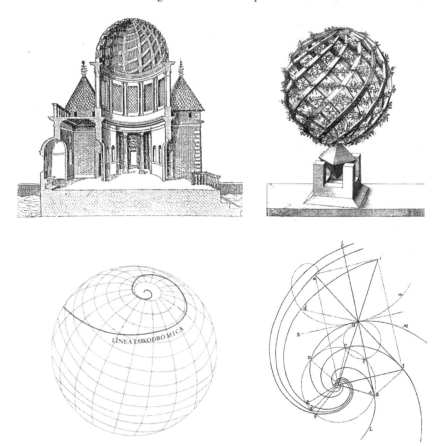

Fig. 29. above, left) De l'Orme, Vault (1561); above, right) Jamnitzer, *Perspective* (1568); below) loxodrome and logarithmic spiral

Perspective, as a science of vision, focused on simple forms whose intuitive character was the origin of the consequent scientific rationalisation derived from mathematical thinking. With the geometric figures related to architecture, in the framework of mutual interest, agreements and exchange occurred between practical aspects and the scientific contributions of mathematicians. In this way, the ribs of de l'Orme's vault at the Château d'Anet were projected on the paving as circumferences, being curves in the vault itself, result of the junction of the cylinder projected by the circumferences and the spherical surfaces. While an architect could have resolved this geometry intuitively, we know that this curve was discovered in the fourth century B.C. by the Greek philosopher and mathematician Eudoxus of Cnidus, who called it 'hipopede', horse-fetter, today it is known as the spherical lemniscate.

Over time, geometry has evolved from intuitive control, based on experimentation, to reach scientific maturity in the mathematical field of abstract reasoning. The geometry related to architecture, as for science in general, almost always had certain empirical roots to become established through the formal language of mathematics. Spirals, helices and helicoids, figures perfectly defined by their properties and equations, are good examples of the complex relations between architecture, aesthetics, science and experimentation, matters that go beyond formal similarities.

The overall conception of the culture gives meaning to the obvious relations between the architectural inventions of de l'Orme, the ornamental forms of Jamnitzer, the study of loxodromes as the curves on the Earth's surface, always directed to the same point of the compass, and the logarithmic spiral, which the mathematician Jakob Bernoulli studied around 1694, dedicating a complete book to it (fig. 29).

Translated from Spanish by Shirley Burgess and Michael van Laake

Notes

1. In 2003, an International symposium, "Pedro Berruguete and his time" was held in Palencia. Many presentations were questioned whether the painter actually spent time in Urbino.
2. *Ceux qui voudront pendre la peine, cognoistront ce que ie dy para la voute spérique, laquelle i'ay faict faire en la Chappelle du Chasteau dÀnet, avecques plusieurs sortes de branches rempantes au contraire l'un de l'autre, & faisant par mesme moyen leurs compartiments qui sont a plombe & perpendicule, dessus le plan & pavé de ladicte Chappelle* [De l'Orme 1567: ch. 11, 122; our trans.].

References

BERCHEZ, J. and F. MARÍAS. 2002. Fray Juan Andrés Ricci de Guevara e la sua architettura teologica. *Annali di architettura* **14**: 251-279.

BLUNT, Anthony. 1958. *Philibert de l'Orme*. London: A. Zwemmer.

DE L'ORME, Philibert. 1561. *Nouvelles inventions pour bien bastir et a petits fraiz*. París.

———. 1567. *Le premier Tome de l'Architecture*. París.

CERCEAU, Jacques Androuet du. 1576-1579. *Le premier volume des plus excellents bastiments de France*. París.

DÜRER, Albrecht. 1525. *Underweysung der Messung*. Nuremberg

HERRERA, J. de. 1589. *Sumario y breve declaración de los diseños y estampas de la fábrica de San Lorenzo el Real del Escorial*. Madrid: Alonso Gómez.

HUYGHE, René. 1965. *Diálogo con el arte*. Barcelona: Labor.

LLAGUNO Y AMIROLA, Eugenio. 1829. *Noticias de los arquitectos y arquitectura de España desde su restauración*. Madrid: Imprenta Real.

LAURENCIO OF SAINT NICHOLAS (Father). 1664. *Arte y Uso de Architectura*. Madrid.

MORRISON, Tessa. 2008. Villalpanda's Sacred Architecture in the Light of Isaac Newton's Commentary. Pp. 79-91 in *Nexus VII: Architecture and Mathematics*, Kim Williams, ed. Torino: Kim Williams Books.

————. 2010. Juan Bautista Villalpando and the Nature and Science of Architectural Drawing. *Nexus Network Journal* **12**, 1: 63-73

PIERO DELLA FRANCESCA. 1984. *De prospectiva pingendi*. G. Nicco-Fasola, ed. Florence: Le Lettere.

SAGREDO, Diego de. 1526. *Medidas del Romano*. Toledo: Remón de Petras.

SERLIO, Sebastiano. 1545. *Libro Secondo, Di Prospettiva*. París

VIATOR [Jean Pelerin]. 1505. *De artificiali perspectiva*. Toul.

VIGNOLA [Jacopo Barozzi]. 1583. *Le due regole della prospettiva pratica*. Rome.

VILLALPANDO, Juan Bautista, and Jerónimo DEL PRADO. 1596-1604. *Ezechielem Explanationes et Apparatus Urbis ac Templi Hieroslymitani*, 3 vols. Rome.

VILLALPANDO, Juan Bautista. 1990. *Juan Bautista Villalpando. El tratado de la arquitectura perfecta en la última vision del profeta Ezequiel*. Spanish edition translated from the original Latin by Fray Luciano Rubio O. S. A. Madrid: COAM.

ZEVI, Bruno. 1973. *Il linguaggio moderno dell'architettura: Guida al codice anticlassico*. Torino: Einaudi. (Spanish translation: *El lenguaje moderno de la arquitectura*. Barcelona: Poseidón, 1978.)

About the author

Lino Cabezas Gelabert has a Ph.D. in Fine Arts; currently he is a professor of the Departamento de Dibujo de la Facultad de Bellas Artes de la Universidad de Barcelona. Previously he was professor at the Escuela Técnica Superior de Arquitectura de la Universidad Politécnica de Cataluña. In these institutions he taught classes of perspective, technical drawing and descriptive geometry. He participated and now collaborates in Ph.D and Master couses, conference cycles and meetings organized in several universities and institutions (Fundación Miró, CaixaForum, MUVIM of Valencia, Fundación Juan March, Museo de Arte Abstracto of Cuenca, Museo Arqueológico of Murcia). Among other subjects he published works about the relationship between art and geometry in journals such as *D'Art* and *EGA Revista de Expresión Gráfica Arquitectónica*. His latest book is *El dibujo como invención. Idear, construir, dibujar* (Madrid: Catedra, 2008).

Chris J. K. Williams

Department of Architecture
& Civil Engineering
University of Bath
Bath BA2 7AY   UK
c.j.k.williams@bath.ac.uk

Keywords: Differential geometry, tiling, tension coefficient, harmonic coordinates, isothermal coordinates, dynamic relaxation

Research

Patterns on a Surface: The Reconciliation of the Circle and the Square

Presented at Nexus 2010: Relationships Between Architecture and Mathematics, Porto, 13-15 June 2010.

Abstract. The theory of heat flow on a surface shows that any curvilinear quadrilateral can be 'tiled' with curvilinear squares of varying size. This paper demonstrates a simple numerical technique for doing this that can also be applied to shapes other than quadrilaterals. In particular, any curvilinear triangle can be tiled with curvilinear equilateral triangles.

Introduction

Fig. 1. The British Museum Great Court Roof by Foster + Partners, Buro Happold and Waagner-Biro

This paper arose out of a re-examination of the way in which the geometry of the British Museum Great Court roof (fig. 1) was derived by defining a single surface in the form $z = f(x, y)$ and then relaxing a grid over the surface [Williams 2000]. Fig. 2 shows the structural grid in dark lines and a finer grid in light lines that was used for the relaxation process. There was no particular requirement that the triangles of the structural grid should be equilateral; other structural and architectural issues were more pressing. Nevertheless it is an interesting question as to whether all the triangles could have been made equilateral.

In the theoretical discussion in this paper we shall assume that we are dealing with a 'fine' grid so that there is little difference between behaviour of the grid and the equivalent continuum as described by classical differential geometry. The Geometric Modelling and Industrial Geometry Research Unit at TU Vienna makes a special study of the discrete differential geometry of coarse grids.

Fig. 2. Great Court Roof structural and 'mathematical' grids

Numerical implementation

It is usual to have a theoretical discussion followed by a description of the numerical implementation. Here we will reverse the order because the numerical implementation is so simple, while the theory is more difficult, at least for those unfamiliar with differential geometry.

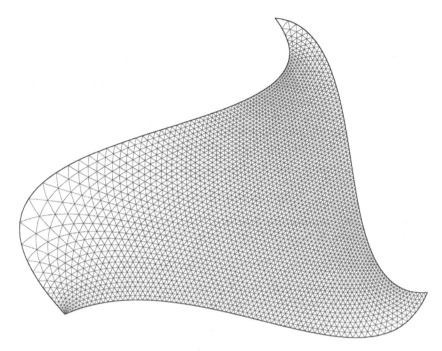

Fig. 3. Curvilinear triangle tiled with curvilinear equilateral triangles

Triangle

Fig. 3 shows a curvilinear triangle tiled with curvilinear equilateral triangles. The triangle is flat so that it can be seen that the tiles are equilateral, but the same procedure can be used if all nodes are constrained to lie on a given curved surface. The figure was produced by simply setting the coordinates of each interior node equal to the average of the coordinates of the six nodes to which it is connected. Each edge node is slid along its boundary curve until the lengths of the projections of the two light lines connected to the edge node onto the boundary itself are of equal length.

Algorithm

It is easier to understand the process if we imagine that the lines on the surface are cables under tension. If the tension in each cable is proportional to its length, then static equilibrium means that the coordinates of each node are the average of those to which it is connected. We can also see that if the nodes are constrained to a surface, all we have to do is to allow the nodes to slide over the surface by removing the component of force in the direction of the normal.

In structural mechanics the tension in a member divided by its length is known as the tension coefficient. In German the word *Kraftdichte* is used, literally, force density. Thus the structural analogy is to use constant tension coefficient cables with nodes that may be constrained to move on a particular surface. Edge nodes are free to slide along the boundary. If the nodes are not constrained to a surface then we shall see that the resulting net forms a minimal surface with a uniform surface tension.

It is possible to make real constant tension coefficient members using coil springs whose coils touch until a certain tension pulls them apart such that the length is proportional to the tension. Such springs were developed by George Carwardine (in Bath) and he used them in the Anglepoise lamp.

The numerical technique finds the equilibrium position by considering the equivalent dynamic problem in which the nodes are moved bit by bit over a large number of cycles. This technique is variously known as dynamic relaxation (invented by Alistair Day), Verlet integration or the semi-implicit Euler, symplectic Euler, semi-explicit Euler, Euler–Cromer or Newton–Størmer–Verlet (NSV) method. The reason for using an iterative technique is that the problem is non-linear unless the nodes are not constrained to move on a surface and the boundaries are straight lines.

Consider a typical internal node, A, whose location is defined by the position vector

$$\mathbf{r}_A = x_A \mathbf{i} + y_A \mathbf{j} + z_A \mathbf{k} \ .$$

$\mathbf{i}, \mathbf{j},$ and $\mathbf{k}$ are unit vectors in the directions of the Cartesian coordinate axes. If it is surrounded by six nodes, B, C, D, E, F, and G, the net force from the six cables is

$$\mathbf{F}_A = \eta(\mathbf{r}_B - \mathbf{r}_A) + \eta(\mathbf{r}_C - \mathbf{r}_A) + \eta(\mathbf{r}_D - \mathbf{r}_A) + \eta(\mathbf{r}_E - \mathbf{r}_A) + \eta(\mathbf{r}_F - \mathbf{r}_A) + \eta(\mathbf{r}_G - \mathbf{r}_A)$$

where η is the constant tension coefficient. If a node is only connected to four nodes then there would only be four contributions to the force.

If the nodes are constrained to move on a surface, $\mathbf{F}_A$ is replaced by

$$\mathbf{F}_A - (\mathbf{F}_A \bullet \mathbf{n})\mathbf{n}$$

in which $\mathbf{n}$ is the unit normal to the surface at A and the $\bullet$ denotes the scalar product. This removes the component of $\mathbf{F}_A$ normal to the surface. It is easiest to specify the surface in the form $f(x, y, z) = 0$, because then the unit normal to the surface is

$$\mathbf{n} = \frac{\nabla f}{\sqrt{\nabla f \bullet \nabla f}} = \frac{\dfrac{\partial f}{\partial x}\mathbf{i} + \dfrac{\partial f}{\partial y}\mathbf{j} + \dfrac{\partial f}{\partial z}\mathbf{k}}{\sqrt{\left(\dfrac{\partial f}{\partial x}\right)^2 + \left(\dfrac{\partial f}{\partial y}\right)^2 + \left(\dfrac{\partial f}{\partial z}\right)^2}} \ .$$

If a node is slightly off the surface it can be put back onto the surface by moving it by

$$-\left(\frac{f}{\nabla f \bullet \nabla f}\right)\nabla f \ .$$

In a time interval δt the velocity of node A changes from $\mathbf{v}_A$ to

$$(1 - \lambda)\mathbf{v}_A + \frac{\mathbf{F}_A}{m}\delta t$$

in which m is the real or fictitious mass of each node and λ is a factor to represent damping. In the same time interval $\mathbf{r}_A$ will change by $\mathbf{v}_A \delta t$.

All the forces on the nodes are calculated in each cycle before updating the velocities and coordinates. The rate of convergence is controlled by the values of λ and $\dfrac{\eta \delta t^2}{m}$, which are both dimensionless ratios. Typically $\lambda = 0.01$ to 0.001 gives the best results and $\dfrac{\eta \delta t^2}{m}$ is chosen by trial and error; if it is too low the procedure is slow, but if it is too large instability will result.

Icosahedron and sphere

The dark lines in fig. 4 are the projection of the edges of an icosahedron onto a sphere. The sphere and icosahedron share the same centre and the projection is done using straight lines. If a plane is covered with a grid of straight lines, it can be projected onto the sphere to form geodesics (fig. 5), again using straight lines through the centre of the sphere. This is known as gnomonic projection and it is almost certainly what Buckminster Fuller used for his geodesic domes. However in fig. 4 the light lines form equilateral triangles and close examination of the figure shows that the light lines do have geodesic curvature – that is, curvature in the plane of the surface – and are therefore not geodesics. It is not possible to have both geodesics and equilateral triangles.

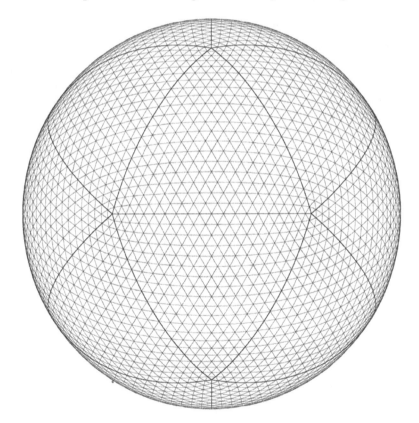

Fig. 4. Icosahedron projected on sphere (dark lines) with equilateral triangle infill (light lines)

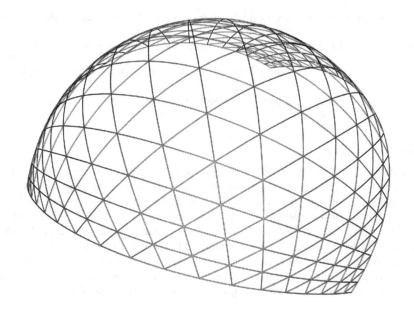

Fig. 5. Three-way geodesics on sphere by gnomonic projection

Hexagon and circle, hexagon and sphere

Figs. 6 a, b and c show a hexagon relaxed onto a flat circle and onto a sphere. On the sphere the grid is repeated twice, once for the top and once for the bottom, the upper and lower parts of figure 6c. The half squares at the edges of figs. 6a and 6b join to form full squares on the sphere.

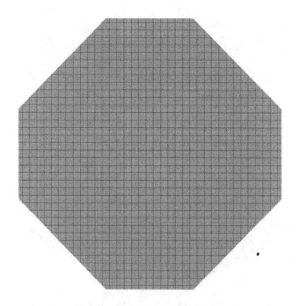

Fig. 6a. Hexagonal grid: cutting pattern

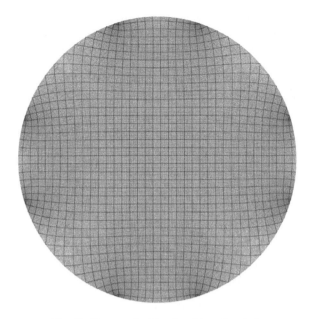

Fig. 6b. Hexagonal grid: relaxed into flat circle

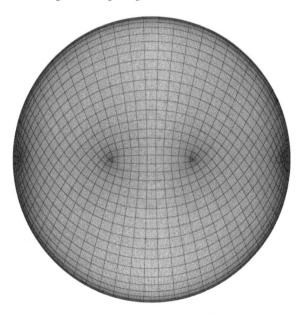

Fig. 6c. Hexagonal grid: relaxed onto sphere

Circle and square

Figs. 7 and 8 attempt the title of this paper, the reconciliation of the circle and the square. There is a clear relationship between fig. 8 and fig. 2, the main difference being that in fig. 2 there is a third set of black lines dividing the quadrilaterals into triangles. The triangles were chosen for the British Museum gridshell primarily for structural reasons. In the numerical work to produce both figs. 7b and 8b it was necessary to automatically adjust the diameter of the circle to achieve curvilinear squares rather than curvilinear rectangles. The reason for this is explained in the theoretical discussion.

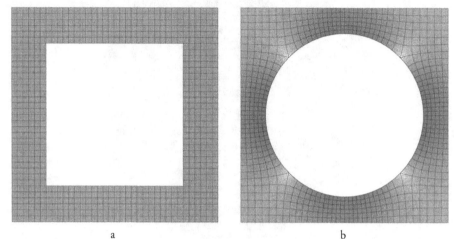

a
b
Fig. 7. Square grid with cut out: (a) cutting pattern and (b) relaxed with central circle

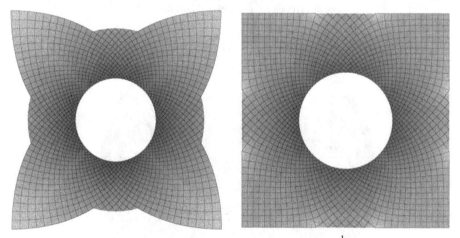

a
b
Fig. 8. Spiral grid: (a) cutting pattern and (b) relaxed with outer square

Frei Otto 'eye'

Finally, figs. 9 and 10 show the Frei Otto 'eye'. Fig. 9 is a physical experiment using washing-up fluid. The trick is to keep the wool loop taut with your fingers while someone else pops the soap film inside the loop. The wool then forms a circle which can be gently pulled up. We will leave discussion of fig. 10 for now, except to say that it was formed in the same way as the other figures with the net automatically forming the minimal surface. Soap film surfaces are minimal because the surface tension automatically reduces the surface area to a minimum.

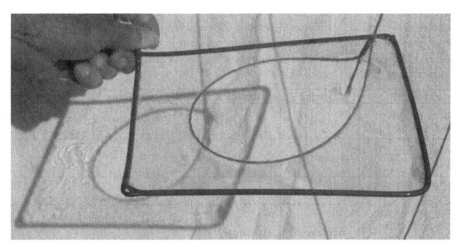

Fig. 9. Frei Otto 'eye' soap film experiment

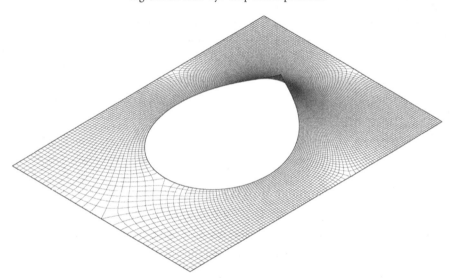

Fig. 10. Frei Otto 'eye' asymptotic lines on the surface forming curvilinear squares

Theoretical discussion

Consider a surface defined by the three equations

$$x = x(u,v)$$
$$x = y(u,v) \text{ or } \mathbf{r} = x(u,v)\mathbf{i} + y(u,v)\mathbf{j} + z(u,v)\mathbf{k}$$
$$z = z(u,v)$$

in which u and v are parameters or surface coordinates and $\mathbf{r}$ is a position vector.

However, we shall not use u and v as parameters, but instead use x^1 and x^2, which are two separate parameters, NOT 'x to the power one' and 'x squared'. The reason for the superscripts is that we can then use the tensor notation. Eisenhart [1940] uses parameters u^1 and u^2, whereas Green and Zerna [1968] use θ^1 and θ^2. Green and Zerna has the advantage that it covers shell theory, that is, the equilibrium of surfaces as well as their geometry. Struik [1988] uses u and v as parameters. Table 1 shows a comparison of the three notations:

Quantity	Struik	Eisenhart	Green and Zerna	This paper
Surface parameters or coordinates	u and v	u^α where α equals 1 or 2	θ^α where α equals 1 or 2	x^i where i equals 1 or 2
Covariant base vectors	$\mathbf{x}_u$ and $\mathbf{x}_v$	$-$	$\mathbf{a}_\alpha = \mathbf{r}_{,\alpha} = \dfrac{\partial \mathbf{r}}{\partial x^\alpha}$	$\mathbf{g}_i = \dfrac{\partial \mathbf{r}}{\partial x^i}$
Contravariant base vectors	$-$	$-$	$\mathbf{a}^\alpha$	$\mathbf{g}^i$
Coefficients of the first fundamental form, components of metric tensor	E, F and G	$g_{\alpha\beta}$	$a_{\alpha\beta}$	g_{ij}
Coefficients of the second fundamental form	e, f and g	$d_{\alpha\beta}$	$b_{\alpha\beta}$	b_{ij}
Christoffel symbols	$-$	$\left\{ \begin{matrix} \lambda \\ \alpha\beta \end{matrix} \right\}$	$\Gamma^\lambda_{\alpha\beta}$	Γ^k_{ij}
Covariant derivative of the components of a vector	$-$	$v^i{}_{,i}$	$v^j\big|_i$	$\nabla_i v^j$
Components of membrane stress in a shell	$-$	$-$	$n^{\alpha\beta}$	σ^{ij}

Table 1

Two way net

In a two way net a typical node (i, j), is connected to four neighbours, $(i+1, j)$, $(i, j+1)$, $(i-1, j)$ and $(i, j-1)$. If the tension coefficient is taken as unity, the resulting out of balance force on node (i, j) is the component of

$$(\mathbf{r}_{i+1,j} - \mathbf{r}_{i,j}) + (\mathbf{r}_{i,j+1} - \mathbf{r}_{i,j}) + (\mathbf{r}_{i-1,j} - \mathbf{r}_{i,j}) + (\mathbf{r}_{i,j-1} - \mathbf{r}_{i,j})$$
$$= (\mathbf{r}_{i+1,j} - 2\mathbf{r}_{i,j} + \mathbf{r}_{i-1,j}) + (\mathbf{r}_{i,j+1} - 2\mathbf{r}_{i,j} + \mathbf{r}_{i,j-1})$$

in the local tangent plane to the surface. The equivalent continuum quantity is the component of

$$\frac{\partial}{\partial x^1}\left(\frac{\partial \mathbf{r}}{\partial x^1}\right) + \frac{\partial}{\partial x^2}\left(\frac{\partial \mathbf{r}}{\partial x^2}\right) = \frac{\partial^2 \mathbf{r}}{\left(\partial x^1\right)^2} + \frac{\partial^2 \mathbf{r}}{\left(\partial x^2\right)^2} = \frac{\partial \mathbf{g}_1}{\partial x^1} + \frac{\partial \mathbf{g}_2}{\partial x^2}$$

in the local tangent plane. Thus for equilibrium,

$$\left(\frac{\partial \mathbf{g}_1}{\partial x^1} + \frac{\partial \mathbf{g}_2}{\partial x^2}\right) \bullet \mathbf{g}_1 = 0$$
$$\left(\frac{\partial \mathbf{g}_1}{\partial x^1} + \frac{\partial \mathbf{g}_2}{\partial x^2}\right) \bullet \mathbf{g}_2 = 0$$

The components of the metric tensor, $\mathbf{g}_{ij} = \mathbf{g}_i \bullet \mathbf{g}_j$, and therefore

$$\frac{\partial g_{jk}}{\partial x^i} = \frac{\partial \mathbf{g}_j}{\partial x^i} \bullet \mathbf{g}_k + \mathbf{g}_j \bullet \frac{\partial \mathbf{g}_k}{\partial x^i}$$

$$\frac{\partial g_{ki}}{\partial x^j} = \frac{\partial \mathbf{g}_k}{\partial x^j} \bullet \mathbf{g}_i + \mathbf{g}_k \bullet \frac{\partial \mathbf{g}_i}{\partial x^j}$$

$$\frac{\partial g_{ij}}{\partial x^k} = \frac{\partial \mathbf{g}_i}{\partial x^k} \bullet \mathbf{g}_j + \mathbf{g}_i \bullet \frac{\partial \mathbf{g}_j}{\partial x^k}$$

$$\frac{\partial \mathbf{g}_i}{\partial x^j} \bullet \mathbf{g}_k = \frac{1}{2}\left(\frac{\partial g_{jk}}{\partial x^i} + \frac{\partial g_{ki}}{\partial x^j} - \frac{\partial g_{ij}}{\partial x^k} \right)$$

since $\dfrac{\partial \mathbf{g}_i}{\partial x^j} = \dfrac{\partial^2 \mathbf{r}}{\partial x^i \partial x^j} = \dfrac{\partial \mathbf{g}_j}{\partial x^i}$.

Thus the equilibrium equations become

$$\frac{1}{2}\left(\frac{\partial g_{1k}}{\partial x^1} + \frac{\partial g_{k1}}{\partial x^1} - \frac{\partial g_{11}}{\partial x^k} \right) + \frac{1}{2}\left(\frac{\partial g_{2k}}{\partial x^2} + \frac{\partial g_{k2}}{\partial x^2} - \frac{\partial g_{22}}{\partial x^k} \right) = 0$$

for $k = 1$ and $k = 2$ The metric tensor is symmetric ($g_{ij} = g_{ji}$) and so

$$\frac{\partial g_{k1}}{\partial x^1} + \frac{1}{2}\frac{\partial g_{11}}{\partial x^k} - \frac{\partial g_{k2}}{\partial x^2} - \frac{1}{2}\frac{\partial g_{22}}{\partial x^k} = 0$$

or

$$\frac{1}{2}\frac{\partial}{\partial x^1}\left(g_{11} - g_{22} \right) + \frac{\partial g_{12}}{\partial x^2} = 0$$

$$-\frac{1}{2}\frac{\partial}{\partial x^2}\left(g_{11} - g_{22} \right) + \frac{\partial g_{12}}{\partial x^1} = 0$$

These are the Cauchy–Riemann equations and produce

$$\frac{\partial^2}{\left(\partial x^1\right)^2}\left(g_{11} - g_{22} \right) + \frac{\partial^2}{\left(\partial x^2\right)^2}\left(g_{11} - g_{22} \right) = 0$$

$$\frac{\partial^2 g_{12}}{\left(\partial x^1\right)^2} + \frac{\partial^2 g_{12}}{\left(\partial x^2\right)^2} = 0$$

Thus both $(g_{11} - g_{22})$ and g_{12} satisfy Laplace's equation. Note that here we have just partial derivatives, NOT covariant derivatives which one would normally associate with differential equations on a surface.

Examination of figures 7a and 7b shows that the sliding boundary condition means that the cables are orthogonal on all boundaries, except at the singular points corresponding to the corners of the inner square. Adjustment of the circle radius removes this problem to produce $g_{12} = 0$ on all boundaries, and therefore Laplace's equation tells us that $g_{12} = 0$ everywhere. Thus $\dfrac{\partial}{\partial x^1}(g_{11} - g_{22}) = 0$ and $\dfrac{\partial}{\partial x^2}(g_{11} - g_{22}) = 0$ so that $(g_{11} - g_{22}) = $ constant everywhere. Thus we must have curvilinear rectangles, all with the same difference in the square of the side lengths. In the case of figure 7b, symmetry tells us that the rectangles must be squares – provided that we have the correct circle radius to ensure $g_{12} = 0$.

The sphere in fig. 6c has no boundaries, although it does have eight poles, whereas the sphere in fig. 4 has twelve poles. In fig. 8b the boundary conditions are mixed. On the circle and towards the middle of each side $(g_{11} - g_{22}) = 0$, whereas towards the ends of each side $g_{12} = 0$. Again it was necessary to adjust the circle diameter to achieve curvilinear squares.

Generalisation

A net with curvilinear squares and constant tension coefficients corresponds to a uniform surface tension equal to the tension coefficient. A uniform surface tension in a general coordinate system corresponds to the membrane stress tensor (see Green and Zerna [1968]) being proportional to the metric tensor, or in components,

$$\sigma^{ij} = \eta g^{ij}$$

in which η is the surface tension.

Let us imagine an initial or reference configuration in which the contravariant components of the metric tensor are G^{ij}. If we want to have a conformal mapping from this configuration to a new configuration, then in the new configuration $g^{ij} = \dfrac{\sqrt{G}}{\sqrt{g}} G^{ij}$

in which $G = G_{11}G_{22} - G_{12}^2$ and $g = g_{11}g_{22} - g_{12}^2$. A conformal map is one which preserves angles and ratios of lengths.

Now if we set $\sigma^{ij} = \dfrac{\sqrt{G}}{\sqrt{g}} G^{ij}$, the equilibrium equations in the local tangent plane of the new surface are

$$0 = \nabla_i \sigma^{im} = \frac{\partial}{\partial x^i}\left(\frac{\sqrt{G}}{\sqrt{g}} G^{im}\right) + \frac{\sqrt{G}}{\sqrt{g}} G^{jm}\Gamma_{ij}^i + \frac{\sqrt{G}}{\sqrt{g}} G^{ij}\Gamma_{ij}^m$$

in which ∇_i is the covariant derivative and $\Gamma_{ij}^m = g^{mk}\dfrac{1}{2}\left(\dfrac{\partial g_{jk}}{\partial x^i} + \dfrac{\partial g_{ki}}{\partial x^j} - \dfrac{\partial g_{ij}}{\partial x^k}\right)$ are the Christoffel symbols. We can use the fact that

$$\Gamma_{ij}^i = g^{ik}\frac{1}{2}\left(\frac{\partial g_{jk}}{\partial x^i} + \frac{\partial g_{ki}}{\partial x^j} - \frac{\partial g_{ij}}{\partial x^k}\right) = g^{ik}\frac{1}{2}\frac{\partial g_{ki}}{\partial x^j} = \frac{1}{\sqrt{g}}\frac{\partial \sqrt{g}}{\partial x^j}$$

to write

$$\frac{\partial}{\partial x^i}\left(\sqrt{G}G^{im}\right)+\frac{\sqrt{G}G^{ij}g^{mk}}{2}\left(\frac{\partial g_{jk}}{\partial x^i}+\frac{\partial g_{ki}}{\partial x^j}+\frac{\partial g_{ij}}{\partial x^k}\right)=0.$$

Finally we can use the Levi-Civita tensor or permutation tensor to write

$$\frac{\partial}{\partial x^i}\left(\sqrt{G}G^{im}\right)+\frac{\sqrt{G}G^{ij}\varepsilon^{mp}\varepsilon^{kq}g_{pq}}{2}\left(2\frac{\partial g_{jk}}{\partial x^i}-\frac{\partial g_{ij}}{\partial x^k}\right)=0$$

in which $\varepsilon^{12}=-\varepsilon^{21}=\dfrac{1}{\sqrt{g}}$, $\varepsilon^{11}=0$ and $\varepsilon^{22}=0$.

If we return to the special case of the curvilinear squares, then $G^{11}=1$, $G^{22}=1$, $G^{12}=0$ and $G=1$, so that

$$2\frac{\partial g_{1k}}{\partial x^1}-\frac{\partial g_{11}}{\partial x^k}+2\frac{\partial g_{2k}}{\partial x^2}-\frac{\partial g_{22}}{\partial x^k}=0$$

which produces

$$\frac{1}{2}\frac{\partial}{\partial x^1}\left(g_{11}-g_{22}\right)+\frac{\partial g_{12}}{\partial x^2}=0$$

$$-\frac{1}{2}\frac{\partial}{\partial x^2}\left(g_{11}-g_{22}\right)+\frac{\partial g_{12}}{\partial x^1}=0$$

as above.

Equilateral triangular nets

If we have a uniform equilateral triangular net in the initial configuration, then $G^{11}=1$, $G^{22}=1$, $G^{12}=-\dfrac{1}{\cos 60°}=-2$ and $G=-\dfrac{1}{1-\cos^2 60°}=\dfrac{4}{3}$ so that

$$2\frac{\partial g_{1k}}{\partial x^1}-\frac{\partial g_{11}}{\partial x^k}+2\frac{\partial g_{2k}}{\partial x^2}-\frac{\partial g_{22}}{\partial x^k}-2\left(2\frac{\partial g_{2k}}{\partial x^1}-\frac{\partial g_{12}}{\partial x^k}+2\frac{\partial g_{12k}}{\partial x^2}-\frac{\partial g_{21}}{\partial x^k}\right)=0$$

or

$$\frac{\partial}{\partial x^1}\left(g_{11}-g_{22}\right)=4\frac{\partial g_{11}}{\partial x^2}-2\frac{\partial g_{12}}{\partial x^2}$$

$$-\frac{\partial}{\partial x^2}\left(g_{11}-g_{22}\right)=4\frac{\partial g_{22}}{\partial x^1}-2\frac{\partial g_{12}}{\partial x^1}.$$

These are satisfied by $g_{11}=g_{22}=2g_{12}$, which is the requirement for a curvilinear equilateral triangle net.

Minimal surfaces

A minimal surface is a surface of minimum surface area which can be physically modelled by a soap film. Minimal surfaces have many interesting properties, amongst which is the fact that the principal curvature trajectories form curvilinear squares on the surface. On a minimal surface the asymptotic directions (which are the directions of zero normal curvature) are at 45° to the principal curvature directions and therefore their trajectories also form curvilinear squares on the surface.

If a soap film has a boundary which is a thread under tension, then equilibrium of the thread dictates that the curvature of the thread must be constant and lie in the local tangent plane to the soap film surface. This means that the boundary must be an asymptotic trajectory. These facts were used to produce the minimal surface in fig. 10. The nodes are free to move in the direction normal to the surface, automatically producing a minimal surface. The main problem is the adjustment of the thread length and the cutting pattern in order to produce curvilinear squares. Included in this adjustment is the fact that the maximum slope along the straight edges must occur at a point where the curvilinear squares are theoretically infinitely large.

Conclusions

The technique provides a relatively simple method of tiling a surface with curvilinear tiles of constant shape, but varying size. In the case of a triangular region equilateral triangles can always be used, but for other shapes care must be taken with the boundary conditions to ensure that the tiles are of the same shape. For minimal surfaces, the technique can both find the shape and produce a principal curvature or, alternatively, an asymptotic direction net.

In this paper we have used the Laplace's equation, or the harmonic equation, associated with the operator ∇^2. This means that we have one boundary condition. The biharmonic equation, associated with ∇^4, allows two boundary conditions, for example the position and slope of a bent elastic plate. The flow of fluids in two dimensions can be modelled using an analogy of a bent plate in which mean curvature of the plate is equal to the fluid vorticity. Fig. 11 was produced this way using techniques not that dissimilar to those employed in this paper. The dark streamlines are the contours of height of the plate and the lighter streak shows the starting of vortex shedding. The reason for including this figure is to demonstrate the complexity of patterns that develop in nature, again effectively from the interaction of rectangular and circular geometries.

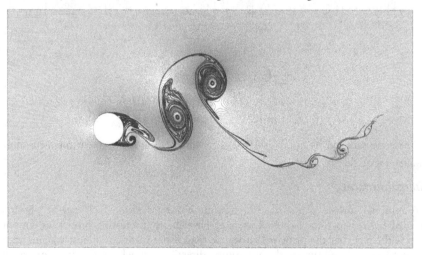

Fig. 11.

References

WILLIAMS, Chris J.K. 2000. The definition of curved geometry for widespan structures. Pp. 41-49 in *Widespan roof structures*, M. Barnes and M. Dickson, eds. London: Thomas Telford.

EISENHART, Luther Pfahler. 1940. *An Introduction to Differential Geometry with Use of the Tensor Calculus*. Princeton: Princeton University Press.

GREEN, A. E. and W. Zerna. 1968. *Theoretical Elasticity*. 2nd ed. Oxford: Oxford University Press.

STRUIK, Dirk J. 1988. *Lectures on Classical Differential Geometry*, 2nd ed. New York: Dover Publications.

About the author

Chris Williams is a structural engineer who worked for Ove Arup and Partners prior to joining the Department of Architecture & Civil Engineering at the University of Bath. He has a particular interest in the relationship between geometrical form and structural action as applied to bridges, shells, tension structures and tall buildings. This leads to the use of specially written computer programs to generate complex, often organic, forms for architectural and structural applications. His work has been applied in practice with architects and engineers including Foster and Partners, Richard Rogers Partnership, Branson Coates Architecture and Buro Happold. His teaching interests include design project work with students of both architecture and engineering, structural analysis, computer programming and continuum mechanics. He has recently lectured at the Architectural Association, London; Graduate School of Architecture, Planning, and Preservation, Columbia University, New York; University of Pennsylvania School of Design, Philadelphia; National Technical University of Athens; Civil Engineering Department, TU Delft, Netherlands; Architecture Department, HTW Chur, Switzerland; Cambridge University Engineering Department; Departmento de Matemática, Universidade de Lisboa, Portugal; Faculdade de Arquitectura da Universidade do Porto, Portugal; Martin Centre, Architectural Research, Cambridge University; Foster and Partners; Arup Associates, London; Fielden Clegg Bradley, London. He is a Smartgeometry Tutor. He was Visiting Professor at the School of Architecture, Royal Academy of Fine Arts, Copenhagen. He has defined the geometry and performed non-linear structural analysis of the British Museum Great Court roof (Foster and Partners, Buro Happold, Waagner-Biro); Weald and Downland Museum gridshell (Edward Cullinan Architects, Buro Happold, Green Oak Carpentry); Savill Gardens gridshell (Glenn Howells Architects, Buro Happold, Green Oak Carpentry).

Marco Giorgio Bevilacqua

Department of Civil Engineering
University of Pisa
Via Diotisalvi, 2
56126 Pisa ITALY
mg.bevilacqua@ing.unipi.it

Keywords: housing, design analysis, design theory, Alexander Klein, minimal surfaces

Alexander Klein (1879 – 1960)

Research

Alexander Klein and the Existenzminimum: A 'Scientific' Approach to Design Techniques

Presented at Nexus 2010: Relationships Between Architecture and Mathematics, Porto, 13-15 June 2010.

Abstract. The urgent need for housing for the working class following World War I led to a search to determine new standards for housing to ensure that minimum requirements for living, or *Existenzminimum*, would be respected while costs were kept low. Overshadowed by more famous architects such as Gropius, Taut, and Le Corbusier, Alexander Klein's important role in this search is often neglected. Klein's remarkably innovative and mathematically rigorous design methodology began with the comparison of various types of dwellings, aimed at the determination of objective terms for the valuation of the design quality. Klein studied the problem of dwelling in its complexity, even considering the effects induced by conditions of living to the human psyche. His points of reference were the needs of the family and the individual, rather than impersonal hygienic-sanitary parameters. Klein's methodological research may still be considered an important point of reference.

Introduction

The figure of architect Alexander Klein is often relegated to the margins of the official historical studies, where he is cited without highlighting the important role his work played in the search for new housing standards during the first half of the twentieth century. Owing to the specific nature of his works and the limited amount of his designs actually built in Europe, his name is overshadowed by the most famous rationalist architects.

Born in Odessa in 1879, Klein took his degree at the St. Petersburg's School of Arts in 1904. In 1920 he decided to move to Germany. Tied to the principles of symmetry, 'equilibrium' and orderliness of the *Beaux Arts* tradition, he remained on the sidelines of the European cultural debate on the Modern Architecture until 1927, when he was nominated a *Baurat*, councilman for the Berlin's town planning. At that point his work was suddenly centered on the necessity to determine, in as short a time as possible, new housing standards for the working class, in order to ensure, at low costs, respect for the minimum requirements for living.

Klein therefore began a long period of study and research on the economic and typological problems related to the development of residential buildings, which led him to carry out assignments of management and research in the *Reichsforschungsgesellschaft für Wirtshaftlichkeit im Bau-und Wohnungswesen* (RFG, the government research agency for the economic and constructive problems of the residential buildings), instituted in 1927 in order to encourage new experiments in model-house projects.

His new theoretical principles were applied in several projects of new urban areas near Berlin – Wilmersdorf in 1927 and Zehlendorf in 1928-29 – and in the project of the *Gross-Siedlung* of Bad-Dürrenberg in Leipzig in 1927.

With the enactment of racial laws in Nazi Germany, owing to his Jewish origin, Klein was forced to leave Germany in 1933, moving to France. Two years later he went to Palestine, where he worked as a teacher at the *Technion*, the Institute of Technology of Haifa, and continued his studies on urban planning; he also designed several projects for new cities.[1]

The urban nature of his projects, his in-depth analysis of the relationship between man and the city, his addressing the problem of housing on scales ranging from that of the individual house to that of an entire urban area, permit us to place Klein in direct comparison with the most famous modern architects who worked on housing, such as Walter Gropius, Bruno Taut, Martin Wagner and Le Corbusier.

Klein died in New York in 1960.

The problem of the house in the Weimar Republic

The second half of the nineteenth century saw a remarkable increase in the population of Berlin, which quickly grew from 170,000 to 1,950,000 inhabitants. In 1920 Berlin was one of the most densely populated cities in the Western world.

At the end of the nineteenth century, the *Polizeiverordnung*, "Regulations of police" enacted in 1853, was the only instrument Berlin had for controlling residential expansion. Its prescriptions were limited exclusively to the minimum dimensions for interior courtyards, based on the overall dimensions of a fireman's wagon, and the maximum height of the buildings in relation to the width of the street. Successive regulations of 1887 and 1897 were integrated into these regulations, introducing hygienic norms for the aeration of rooms and limiting the use of basements as spaces for living.

In working class quarters, high rent for houses, often reaching a value equivalent to the 40% of a medium worker's salary, forced two or more families to share a single house (fig. 1).

Fig. 1. Berliner Mietskasernen. Exterior and interior views of tenement building development in the early years of the twentieth century

During the first world war the situation was aggravated by a paralysis in the construction of new housing. At the end of the war in Germany, the problem of housing shortages for the proletariat – exacerbated by the precarious economic situation, the shortage of labour, inflation and the consequent increase in the price of raw materials – reached a level never before seen.

Finally, with the Republic of Weimar, the social-democratic government decided to grapple seriously with the problem of the housing requirements. After several transitory provisions, aimed at controlling the prices of building materials and rents, in 1924 the government, encouraged by the stabilization of the mark and the influx of foreign capital, began to promote interventions on a vast scale.

The resumption of building and public financing for mass-production housing favored the formation of the cooperative societies, which in those years realized the majority of the constructions.[2]

Building production was strictly tied to the decisions of the public administration, which gave power to some "technicians" in the field of town planning: the role of the *Staudtbauten*, architect-administrator, was a significant one in politics. This was Klein's role in the public administration and in the RFG.

In 1932 the advent of National Socialism put a stop to any form of initiative, causing more serious problems and hurling Europe towards a second world war.

Alexander Klein and rationalism

The deep economic crisis into which all European nations fell after World War I fostered a sense of deep social and political responsibility in architects: architecture was viewed as a social service. It was clear that it was necessary to go beyond the cultural and social-bourgeois tradition of the late nineteenth century, which had begun to show its inadequacy to face the new problems that went hand-in-hand with scientific, economic and industrial progress.

From this point of view, the rationalist movement in architecture developed autonomously with respect to the artistic vanguards: starting from the principle of "art for everybody", it elaborated a planning method that placed architecture in service of the society, recognizing in it the capacity to impact the political order deeply and bring about the resolution of social problems. The problem of housing requirements for the working class was a priority. In his 1934 *Das Einfamilienhaus – Südtyp. Studien und Entwürfe mit grundsätzlichen Betrachtungen*, Alexander Klein wrote:

> Every age has its great architectonic themes; in ancient times they were above all temples and public buildings, in the Middle Ages churches and castles, in the Renaissance villas and palaces, then bourgeois mansions, and today the mass-production houses, together with the industrial buildings. As in the past dominant ideas found an adequate architectural expression, so today the mass-production house, with its economic-social requirements, cannot help but influence the architectonic expression [Baffa Rivolta and Rossari 1975: 141].

In regards to this, Giuseppe Samonà wrote:

> The rationalist architects saw the house as an ethical symbol, and at the same time it led them to act with logical rigour. The house and the

neighborhood were at the center of the moral exigency ... to reach, with the coherence between function and form, a harmony that operated from inside the "cell" in which man lives, indicating a way to solve all social conflicts [Samonà 1959: 83].

The faith in "logic" distinguished all the rationalist architects: in the attempt to plan architectural spaces as rationally as possible, all exigencies were definable as a measurable requirement; for any given exigency, architects could propose a solution considered as the most advantageous and profitable from many points of view.

The determination of the *Existenzminimum*, which was the focus of the intense activities of Gropius, May and Klein, started by dimensioning the habitation cell on the basis of what was necessary to satisfy men's exigencies, in a socialist vision that considered all men equal, regardless of their social class.

This task led to the formulation of universal standards for building, which also served to promote the process of industrialization, "the outgrowth of all the rationalist technique's logic, that is, obtaining the maximum social result with the minimum economic effort" [De Fusco 1982: 219].

Le Corbusier wrote in this regard:

> We must see to the establishment of *standards* so we can face up to the problem of *perfection*. ... Architecture works on standards. Standards are a matter of logic, of analysis, of scrupulous study: they are based on a problem well posed. Experimentation definitely fixes the standard. ... To establish a standard is to exhaust all the practical and reasonable possibilities, to deduce a recognized type consistent with function, maximal return with minimum expenditure of means, manpower, and materials, words, forms, colors, sounds [Le Corbusier 2008: 182-186].

Having defined the typological characters of the habitation cells, attention was extended from the individual dwelling to the building, as a combination of the several units based on the same conforming principles; several buildings, arranged on the basis of principles of orientation, means of communication and infrastructures, determined the neighborhood; several neighborhoods constituted the city. A feeling of optimism deeply permeated the rationalist architects' treatises; Klein wrote: "we think that a city conceived in this way and based on scientific and human principles will help the community to form better citizens who will consequently create a better community" [Klein 1947; cited in Baffi Rivolta and Rossari 1975: 184].

The method of analysis in the minimum design of a lodging

Klein's works on *Kleinwohnungen*, "little lodgings", were shown in 1928 at the Housing Exposition, on the occasion of International Congress for Housing and Planning in Paris.

The works, which proposed a comparison between models for several houses, aimed at determining objective terms for the evaluation of a design's quality, in order to find what arrangement best defined minimum lodging.

At the time when Klein exhibited his works, the Berlin building regulation prescribed only a few limitations on the areas of rooms for the mass-production houses, leading to excessively large areas for the houses. The consequent high building costs meant that they were too expensive for the working class.[3]

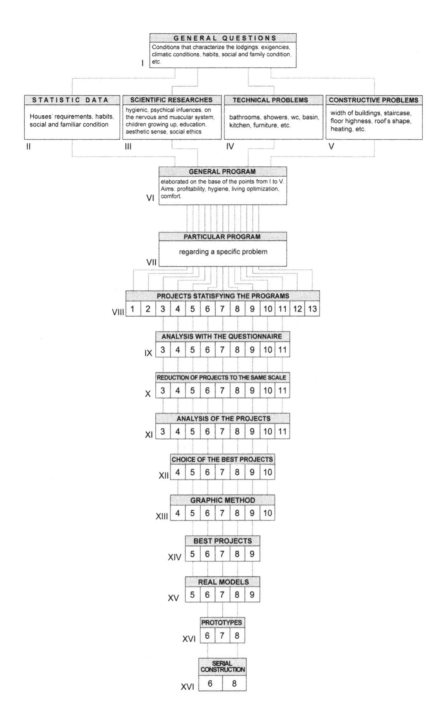

Fig. 2. General scheme of the method for planning rational residential typologies

Fig. 3 table — "score method" applied to nine models of lodgings.

11	10	9	8	7	6	5	4	3	Nr.	PROJECT NR.		
65.76	75.46	73.14	89.40	90.00	80.94	88.00	66.82	74.18	1	BUILT AREA	MAIN CHARACTERS OF THE HOUSE	ECONOMIC ASPECTS
274	324	314	393	387	360	392	283	319	2	BUILT VOLUME		
51.47	53.18	64.45	63.02	65.45	61.33	65.90	51.10	58.35	3	USED AREA		
2	1.5	2.5	2.5	2.5	2.5	2.5	2	2	4	AMOUNT OF ROOMS		
2	3	3	4	3	3	2.5	2	2	5	AMOUNT OF BEDS		
32.90	25.15	29.76	22.26	30.00	26.98	29.33	33.41	37.09	6	BETTEFFEKT: 1/5		
137	108	128	98	129	120	131	141	159	7	BUILT VOLUME PER BED 2/5		
16.40		20.60	20.00	17.30	20.00	21.00	20.25	24.10	8	LIVING ROOM'S AREA	MAIN ROOMS	
13.50	32.60	22.60	26.45	28.95	22.75	24.40	14.00	13.90	9	BEDROOM'S AREA		
29.90	32.60	43.20	46.45	46.25	42.75	45.40	34.25	38.00	10	RESULTING AREA 8+9		
15.30	11.40	11.30	8.70	10.60	10.80	11.20	9.80	9.75	11	KITCHEN'S AREA	SECONDARY ROOMS	
3.85	4.32	4.75	3.92	4.30	4.00	5.00	3.65	4.20	12	BATHROOM'S AREA		
2.42	4.86	5.20	3.95	4.30	3.78	4.30	3.40	6.40	13	SERVICE AREA		
21.57	20.58	21.25	16.57	19.20	18.58	20.50	16.85	20.35	14	RESULTING SERVICE AREA 11+12+13		
0.783	0.705	0.722	0.705	0.727	0.758	0.749	0.765	0.787	15	NUTZEFFEKT: 3/1	COEFF.	
0.455	0.432	0.483	0.520	0.514	0.528	0.516	0.512	0.512	16	WOHNEFFEKT: (8+9)/1		
+	+	-	+	-	-	-	+	+	17	Is the orientation homogeneous both in the living and in the bedrooms?	HYGIENIC ASPECTS	
-	-	-	+	-	+	-	-	+	18	shadows avoided in living and bedrooms ?		
-	-	-	+	-	+	-	-	+	19	Is the light sufficient ?		
+	-	+	+	+	+	+	+	+	20	not-served rooms avoided?	CHARACTERISTICS REGARDING THE HABITABILITY	
-	-	-	-	-	-	-	-	-	21	may children be divided in base of the their sex ?		
-	-	-	-	-	-	-	-	-	22	Is the rooms' dislocation good for the habitibility?		
-	-	-	-	-	-	-	+	-	23	Is the bathroom separated from the toilette?		
-	+	+	+	+	+	-	-	+	24	Is the access to the loggia indipendent from bedrooms?		
+	+	-	-	-	+	-	-	-	25	Is the position of doors and windows good for the furniture's disposition?		
-	-	-	-	-	-	-	-	-	26	Are bathroom and w.c. adjacent to bedrooms and indipendent of them?		
-	-	-	-	-	-	-	-	-	27	Are there spaces for wardrobes?		
+	-	-	-	-	+	-	-	+	28	Are movement areas concentrated?		
-	-	+	+	+	+	-	+	-	29	Are rooms differentiated in base of use and dimensions?	SPATIAL AND DISTRIBUTIVE CHARACTERISTICS	
+	-	-	+	-	+	+	+	-	30	Disadvantageous connections between rooms avoided?		
-	-	-	-	-	-	-	-	-	31	Are rooms well connected?		
-	-	-	-	-	+	-	-	+	32	Is the light aesthetically good?		
-	-	-	-	-	-	-	-	-	33	Are encumbrances reduced using wall-wardrobes?		
+5	+3	+3	+7	+3	+9	+2	+5	+7		SCORE		

Fig. 3. Example of application of the "score method" to nine models of lodgings

Klein singled out the limitations of the Berlin regulations in considering the surface area as a unitary reference parameter, which reflected the permanence of traditional models both for the determination of spaces and for the arrangement of furniture.

Gropius had defined *Existenzminimum* as the "minimum of space, air, light and heat necessary to men for developing their own vital functions without restrictions due to the lodging" [Gropius 1959: 126]. Klein added psychological aims as well: "we must keep in mind the influence of symptoms of psychological fatigue that negatively influences man's nervous system, caused by unpleasant feelings generated by an accidental disposition of the elements of the plan" [Baffa Rivolta and Rossari 1975: 40].

Klein set forth the problem of housing in all its complexity, enriching the analysis with parameters for evaluating the psychological effects on men. Thus he referred the lodging's dimensions, not to the areas of the rooms, but to the number of beds contained, that is, to the family to which the lodging was assigned.

The number of beds became the unit of measurement for all the exigencies of living, and consequently determined the amount of space needed for the living and dining rooms, kitchen and bathrooms.

Fig. 2 shows the levels and phases of analysis necessary for to determine the rational types of lodgings according to the model proposed by Klein.

In the first phase, the research was carried out without regard to problems of construction or building materials, in order to limit the amount of variables in this level.

Once the general problems regarding the geographic and cultural context (localization of lodgings, climatic conditions, local customs, familiar and social conditions, etc.) had been brought into focus, and the statistical data on the requirement for actual lodgings and the income of the population collected, the program then continued with in-depth research on the influences of the lodging on the inhabitants (influences related to hygiene, psychology, influences on the nervous and muscular systems, the increase in the number of children, education, the aesthetic sense, the social ethic). This level continued with the analysis of different possibilities for the design of the technological system, followed by the research on problems of construction (depth of the building, amount of flats for each staircase, height of the first floor, organization of the basement, shape of the roof, central or individual heating system). The result of this phase was the formulation of the "General Program".

The "General Program" can be synthesized in four points:

1. the lodging must be inexpensive and every unitary parameter that defines the lodging's value must be minimized;[4]

2. the lodging must be hygienic, that is, complete with sanitary fixtures and rooms that are adequately sunny and ventilated;

3. the lodging must be without defects; its arrangement must guarantee that family life is carried out in the best conditions, making it possible to do different tasks without spending too much energy and time;[5]

4. the lodging must invoke a pleasant impression of spaciousness, with spaces harmonizing in form, light and colors.[6]

Klein maintained the necessity of involving different areas of knowledge in the elaboration of the general program, which are those of the architect, the public agencies and all the various scientific and technical experts, anticipating what came to be the established conviction that architectural planning demands the collaboration of various professional figures.

After the elaboration of the general program, it was the exclusive responsibility of the architect to plan the typologies that efficiently satisfied all the highlighted requirements while keeping costs low. At this point, the program formulated a sort of questionnaire for evaluating the designs in relation with the quality of living provided. Klein introduced the "Score Method" (fig. 3): a certain number of characteristics were fixed on the basis of the general program, so the comparison between models for different lodgings was possible, assigning a positive or a negative score according to their correspondence to the requirements. The models that achieved a mainly positive score were obviously preferred to the others. Understanding that objectivity depended on the choice of the questions and the importance attributed to any of these, Klein introduced some corrective coefficients, in order to confer objectivity to his method. The coefficients were: the *Betteffekt* or "bed effect", the ratio between the total built area and the number of beds; the *Nutzeffekt*, the ratio between the used area and the total built area; the *Wohneffekt*, the ratio between the area of the living room, the bedrooms and the total built area.[7]

Following this phase, there was one called the "Method of the Successive Increments". Here the plans previously selected were grouped on the basis of some dimensional parameters and the distributive scheme, so as to be "reduced to the same scale", that is, to be comparable on the basis of the number of beds. The planimetric diagrams were therefore modified by increasing the length and the width of the building by constant amounts; they were disposed in a grid, where the rows represented the increase in depth, the columns the increase of the width (fig. 4). The result was that the best plans, both in terms of profitability and habitability, were those located along the diagonal of the grid. The plans in the upper half of the grid were neither profitable, hygienic nor practical, while those placed in the lower half were hygienic but not profitable, due to their excessive length.

The third phase was the "Graphic Method" (fig. 5). Klein thought that this was the most reliable and rigorous of all the "methods", because it was more objective and not very susceptible to personal interpretations. The graphic method allowed numerous analyses to be made on the selected plans, making it possible to see the internal connections, the shape of the area engaged by the movements, the space not occupied by the furniture, the area of shadow projected by the walls and the furniture when caught by the sunlight. Designing the interior elevations of any single space, it was also possible to evaluate the correct use of the internal walls and the effects of the position of furniture and openings on the user's psychological conditions. The use of these parameters in the comparison of the selected solutions made it possible to identify the optimal residential typology that satisfied all the requirements.

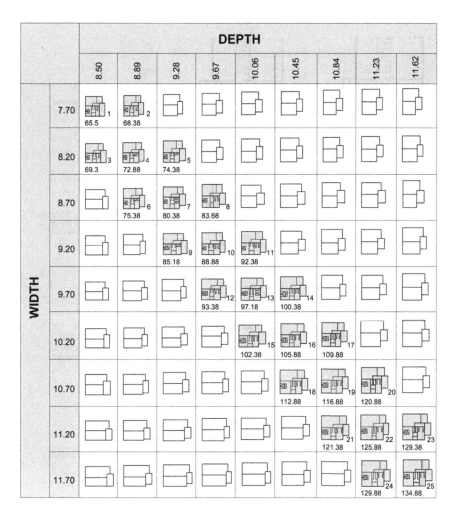

Fig. 4. Method of the successive increments. Example of comparison and evaluation of several plan diagrams reduced to the same scale (1 living room, 1 bedroom for parents, 1 bedroom for children)

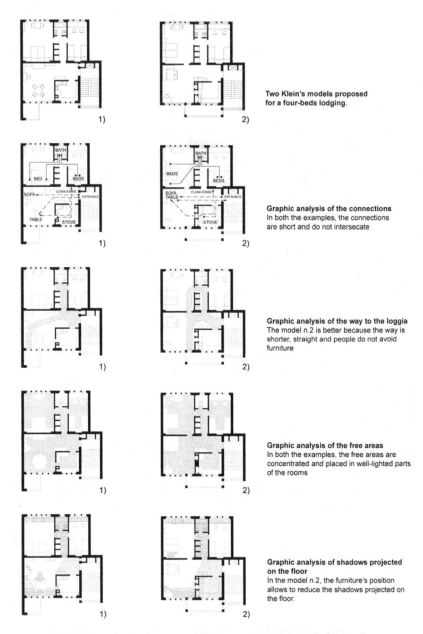

Two Klein's models proposed for a four-beds lodging.

Graphic analysis of the connections
In both the examples, the connections are short and do not intersecate

Graphic analysis of the way to the loggia
The model n.2 is better because the way is shorter, straight and people do not avoid furniture

Graphic analysis of the free areas
In both the examples, the free areas are concentrated and placed in well-lighted parts of the rooms

Graphic analysis of shadows projected on the floor
In the model n.2, the furniture's position allows to reduce the shadows projected on the floor.

Fig. 5. Graphic method. Comparison between two plan models for four beds, published by Alexander Klein in "*Grundrissbildung und Raumgestaltung von Kleinwohnungen und neue Auswertungsmethoden*", Berlin 1928

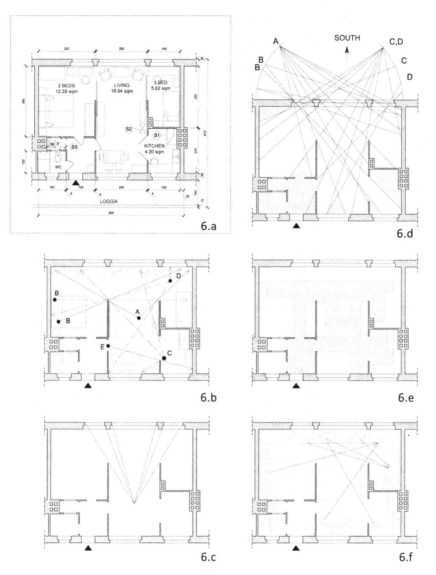

Fig. 6. Example of minimal lodging, planned by Alexander Klein for a loggia house typology (type IIB) and published in the "*Zu dem zusätlichen Bauprogramm der Reichsregierung*", Berlin 1931

The project for a "loggia house" typology, published by Klein in "*Zu dem zusätlichen Bauprogramm der Reichsregierung*" in 1931 (fig. 6) clearly shows how the author contrived some expedients in order to limit the disadvantageous effects of the minimal area. These can be synthesized as:

- wide sliding doors make it possible to extend the air-volume of the bedrooms into the living room (fig. 6a);
- the heating system is differentiated on the basis of the seasons: one coal stove for mid-seasons (S1), one stove in the living room for the normal winter days (S2), an additional stove (S3) for the coldest days (fig. 6a).
- having wide views from every point of the lodging makes it possible to avoid the unpleasant impression of too-narrow spaces (fig. 6b);
- the interior space is placed in visual connection with the exterior environment by using wide sliding doors placed orthogonal to the wall of the windows (fig. 6c);
- the correct positioning and dimensioning of windows allow sunlight to enter all areas of the house (fig. 6d);
- the layout of the space and the arrangement of the furniture are conceived so that the areas left free are concentrated and compact (fig. 6e);
- the interior views are conceived in order to make it easy for parents to watch their children (fig. 6f);
- the glass door between the kitchen and the living room, together with careful arrangement of the furniture, allows mothers to watch their children while they cook (fig. 6g);

The lodging appears to have designed with the woman of the house in mind. Klein recognized how hard housework is, and conceived the house in order to allow women to carry out their household duties easily, with less waste of energy and time, while at the same time being able to watch over their children, in a welcoming and sunny house.

The Gross-Siedlung of Bad Dürrenberg in Leipzig

Alexander Klein was the designer of many projects ranging in scale from the single building to entire urban areas, but just few of them were realized. A chronology of Klein's projects is given in the Appendix.

In the project for the Gross-Siedlung at Bad Dürrenberg in Leipzig in 1930 (fig. 7), Klein finally had the opportunity to test the design method theorized in the years before. The program called for 1000 lodgings laid out in a plot of land 185,769 square meters wide, with an incidence of approximately one lodging each 200 square meters, subdivided in (fig. 8):

1 Loggia house typology
- 187 apartments with 2 ½ beds (type L);
2 Apartment blocks
- 420 apartments with 3 ½ beds (type C 2);
- 118 apartments with 4 ½ beds, with the possibility to separate children on the basis of gender (type C7);
- 48 apartments with 5 ½ beds (type C 13);
3 Single family houses
- 120 houses with 4 ½ beds, without the possibility to separate children on the basis of gender (type E1);
- 21 houses with 5 ½ beds (type E2).

Fig. 7. The Gross-Siedlung of Bad Dürrenberg in Leipzig. View of the area

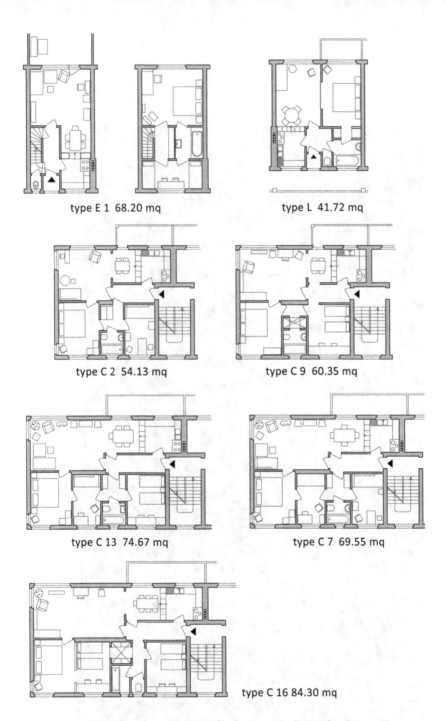

type E 1 68.20 mq   type L 41.72 mq

type C 2 54.13 mq   type C 9 60.35 mq

type C 13 74.67 mq   type C 7 69.55 mq

type C 16 84.30 mq

Fig. 8. Plans of typological lodgings for the Gross-Siedlung of Bad Dürrenberg

All the blocks are arrayed along the north-south axis, in order to have the rooms destined for daytime use exposed to the west and those destined for nighttime use exposed the east. Only in the loggia types are blocks arrayed along the east-west axis, with all the service rooms exposed to the north and the living spaces opened to the south.

Blocks of house with two floors are situated orthogonally to the loggia types and are separated from the main road – the Lützenerstrasse – by a wide, triangular-shaped open space, with sport areas and playgrounds.

In the eastern corner of the area, the square where the station is located represents the main traffic node and is, at the same time, the terminal of the tramway connecting to the Leuna Factory nearby. The single family houses, with two or three living floors, are located near the square of the station.

The main axis of the whole area is marked by a wide, linear park laid out in an east-west direction, with stretches of water, pedestrian ways, trees and benches. At its eastern end, rows of poplars separate the central green park from the blocks of three-floor apartments; playgrounds and sand boxes are provided between the blocks. Roads for cars are relegated to the boundary of the area, and terminate in little squares for the u-turns; only narrow pedestrian pathways – ranging in width from 1.60 to 2.25 meters – arrive to the entrances to the blocks. Places for garbage collection are located between the blocks and the boundary streets.

Conclusions

Unfortunately, the results of Klein's studies have been often disregarded in the rush to build speculatively, to increase the building contractors' profits and earnings from sales of real estate. In neglecting the social ideals of the *Existenzminimum*, the mass-production lodgings failed to provide "the function of shelter from the contradictions and the conflicts of the city ... privileged place for the privacy, rest and the regeneration of the labourer" [Baffa Rivolta and Rossari 1975: 40]. Given the choice between Gropius's "minimum vivendi" and "modus non moriendi", that is, between the minimum conditions for an optimal quality of life and the limit conditions for survival, the latter was often preferred.

However, Alexander Klein's lessons may be considered as universal and continue to play an important role in architecture. According with De Carlo, the fact that architecture then went beyond the methodological bounds of Klein, and of rationalism in general, is proof of the validity of a method "that made the research concrete and, at the same time, made it possible to go beyond it" [De Carlo 1965: 324]. Klein's scientific rigor has not be interpreted as rigidity of his method, but forms an important foundation for the affirmation of a culture, and helped lead architecture out from the dogmatic reiterance of traditional models.

Appendix: Klein's project list

1909-1916 Peter the Great Hospital, Petersburg;

1913-1914 Apartments in Petersburg;

1922-1925 Houses for rent in Berlin-Wilmersdorf;

1925 Houses in Tempelhofer Felde, Berlin (project);

1925 Houses for the Moscow Soviet Price (project);

1926 Five residential buildings and cotton factory in Iwanowo Wosnessensk (project);

1926 Terrace in Berlin-Dahlem (project);

1926-27 Residential building in Berlin-Wilmersdof (project);

1927 Gross Siedlung Bad Durrenberg a Merseburg, Leipzig;

1927 Schultz House in Berlin-Dahlem;

1928 GAGFAH Siedlung Onkel Tom's Hutte in Berlin-Zehlendorf (urban plan);

1928 three buildings in the GAGFA Siedlung Fischtalgrund in Berlin-Zehlendorf;

1928 Lodge for the Heim und Tecnick Competition in Munchen (project);

1929 Klinke House in Berlin-Dahlem;

1930 Residential building at the Kornerpark in Berlin-Neukoll;

1931 Apartments for single workers in Berlin (project);

1931 Enlargeable house for unoccupied workers (project);

1932 House for the Zuruck zum Haushalt in Berlin-zoo (project);

1938-47 Residential quarter near Tiberias (project);

1938-50 Residential quarter Kiriat Yam, Haifa;

1940-50 Residential quarter Tivon, Haifa (project);

1940-50 Residential quarter Mount Carmel-Rushmiah, Haifa (project);

1953-54 Urban Plan for the Technion Campus, Haifa;

1957 Residential building for the Interbau Exposition (project);

Notes

1. Residential quarter near Tiberias (1938-1947); residential quarter *Kiriat Yam*, Bay of Haifa (1938-1950); residential quarter *Tivon* near Haifa (1940-1950); residential quarter *Mount Carmel-Rushmiah* near Haifa (1940-1950); urban plan of the Technion's campus, Mount Carmel near Haifa (1953-1954); see the Appendix for a chronology of Klein's projects.
2. The most important cooperative societies were the GEHAG and the GAGFAH. Martin Wagner and Bruno Taut worked for the GEHAG; Heinrich Tessenow worked for the GAGFAH. The public administration obliged the cooperative societies to respect rules for the dimensioning of spaces in order to be entitled to public subventions and to fiscal privileges [Baffa Rivolta and Rossari 1975: 13].
3. A house had to be provided with a room of at least 20 square meters; for the other rooms the minimum area had to be equal to 14 square meters, 10 square meters for the kitchen, 6 square meters for the *Kammer*, an unspecified room that could be used either as a storeroom or as single bedroom. Thus, the mean area was 100 square meters for a three-room house and 130 square meters for a four-room house, with costs fluctuating between 12,500 and 16,000 marks, urbanization costs excluded.
4. The unitary parameters identified by Klein were: the ratio between the total built area and the number of beds, the ratio between the sum of living-room's areas with the bedrooms' ones and the total built area.
5. "...the amount of rooms must correspond to the family composition without sub-renters ... the bedroom for parents must be separated from children's rooms; it should be possible to separate children on the basis of gender; living-room must be separated from bedrooms;

kitchen must be separated from dining-room … doors and windows must be arranged for leaving sufficient space for placing the indispensable furniture", [Klein 1928], quoted in [Baffa Rivolta and Rossari 1975: 82]. The fact that Klein felt the necessity to emphasize some requirements that currently belong to the residential building's planning, evidences the deep state of precariousness of the German mass-production housing in the first post-war period and marks the importance of his works for our architectonic culture.

6. This is the concept of the "calm house", that is, the importance that the perception of the spaces does not induce unpleasant and anxious feelings in the residents.

7. From the study of the questionnaires, the corrective coefficients do not seem to affect the evaluation by means of an algorithm that justifies the formal rigor. Instead it is more likely that the corrective terms were used by Klein in order to estimate the validity of the judgment on the basis of unavoidably subjective considerations. This limit was however clear to Klein, who admitted that a certain objectivity was only possible in the successive phase of the graphic method.

Bibliography

BAFFA RIVOLTA, M. and A. ROSSARI, eds. 1975. Alexander Klein. Lo studio delle piante e la progettazione degli spazi negli alloggi minimi. Scritti e progetti dal 1906 al 1957. Milano: Mazzotta.

SAMONÀ, G. 1959. L'urbanistica e l'avvenire della città. Bari: Laterza.

DE FUSCO, R. 1982. Storia dell'architettura contemporanea. Napoli: Laterza.

———. 1999. *Mille anni di architettura in Europa.* Bari: Laterza.

LE CORBUSIER [Pierre Jeaneret]. 2008. *Towards an Architecture,* John Goodman, trans. London: Frances Lincoln Ltd. (1st ed. *Vers une architecture,* Paris: Fréal & C., 1923).

GRIFFINI, E. A. 1933. *Costruzione razionale della casa.* Milano: Hoepli.

DE CARLO, G.C. 1965. "Funzione delle residenze nella città contemporanea". *Questioni di architettura e di urbanistica,* Urbino: Argalia.

GROPIUS, W. 1959. Die soziologischen Grundlagen der Minimalwohnung. Die Wohnung fur das Existenzminimum. Frankfurt. (from the Italian translation in Architettura Integrata. 1959 Milano: Mondadori).

Klein, A. 1928. Grundrissibildung und Raumgestaltung von Kleinwohnungen und neue Auswertungsmethoden. *Zentralblatt der Bauverwaltung* **48**: Berlin

———. 1931. Zu dem zusätlichen Bauprogramm der Reichsregierung. *Gesundheitsingenieur* **4**. Berlin.

———. 1934. *Das Einfamilienhaus – Südtyp. Studien und Entwurfe mit grundsätzlichen Betrachtungen.* Stuttgart: Hoffman.

———. 1947. Man and town. In *Yearbook of the American Technion Society,* New York.

About the author

Marco Giorgio Bevilacqua is an Italian engineer living and working in Pisa. He received his degree in civil engineering from the University of Pisa, where he also earned his Ph.D., with research concerning the architectural survey of the fortifications built in Pisa in the sixteenth century. He teaches architectural drawing in the Faculty of Engineering at the University of Pisa. He also teaches architectural design at the UFO University of Tirane (Albania) and digital drawing at the Naval Academy of Lebourn (Italy). He has participated in various national and international projects with research in the field of architectural surveying of monuments and urban landscapes. In the same area, he has published many articles in journals and books and has presented papers at numerous national and international conferences. He is the author of "The conception of Ramparts in the Sixteenth Century: Architecture, Mathematics and Urban Design" (*Nexus Network Journal* vol. 9, no. 2, Autumn 2007) and, with Roberto Castiglia, "The Turkish baths in Elbasan, Albania" (*Nexus Network Journal* vol.10, no. 2, Autumn 2008).

João Paulo Cabeleira
Marques Coelho

Escola de Arquitectura da
Universidade do Minho
Campus de Azurém
4800-058 Guimarães PORTUGAL
joaocoelho@arquitectura.
uminho.pt

Keywords: Inácio Vieira,
perspective, optics, history of
science, quadratura, Baroque
architecture, treatises

Research

Inácio Vieira: Optics and Perspective as Instruments towards a Sensitive Space

Presented at Nexus 2010: Relationships Between Architecture
and Mathematics, Porto, 13-15 June 2010.

Abstract. The manuscripts of Jesuit mathematician Inácio
Vieira (1678-1739) played a significant rule in treatise
production in Portugal. The *Tractado da Óptica* (1714) and
Tractado de Prospectiva (1716) address the nature and
properties of vision, the deception and disillusion of seeing
and the fundaments of perspective. Geometric properties of
visual rays, principles of optical illusion and values inherent
in the optical and mathematical representation of space are
explored, together with practical applications in architecture
and painting. Although he does not make an original
scientific contribution, he formulates operative statements
applied to the arrangement of a sensitive space. This approach
to optics and perspective brings up the relationship between
reality and appearance in which the production of an image
lies in the fact that to show an object, one has to show it *as it
is not*. Vieira's work forges links between optics (*perspectiva
naturalis*) and perspective (*perspectiva artificialis*), analyzing
image distortion through spatial perception, illusion and
quadratura. Pursuing an empirical architecture, the simulated
space of *quadratura* is understood as a challenge that creates a
new dimension and experience of architectural space. Form
and metric properties, from real and illusory construction, are
interrelated contributing to an imaginary perception of
spatiality. It is here that optics and perspective are established
as architectural instruments for achieving a sensitive space.

Introduction

This present study examines optics and perspective treatises, searching for
intersections between science and architectural design processes, seeking *imaginary
architectures*. The connection between perspective theory and practical application
depicts the interplay between built space (real) and represented space (illusory), pursuing
an empirical architecture based upon simulations of *quadratura*, a technique of
illusionary painting.[1] The possibility of an unlimited space beyond architectural form is
given by perspective, not based on the construction of a tectonic reality, but on the
appearance of two-dimensional architectural representation images converted to three-
dimensional illusory architectural spaces. Form and metric properties, of both real and
illusory constructions, are interrelated and contribute to the perception of an imaginary
spatiality. It is then that optics and perspective establish themselves as architectural
instruments and include spatial-visual perception in architectural conception.

Here we focus on the work of the Jesuit Priest Inácio Vieira, professor at *Aula da
Sphera* (Sphere Class) in Lisbon, and his treatises on optics and perspective. *Tractado da
Óptica* (1714) and *Tractado de Prospectiva* (1716). Vieira was also the author of a

number of other treatises on subjects such as astronomy, pyrotechnic mathematics, catoptrics, dioptrics, hydrography (the art of navigation), chiromancy and astrology.

The Aula da Sphera

From the sixteenth to the eighteenth century, the *Aula da Sphera* of the Jesuit College in Lisbon played a major role as the main scientific institution in Portugal, ensuring education on physical-mathematical subjects and integrating scientific breakthroughs by hosting teachers from colleges throughout Europe.

The opening of the Jesuit colleges in Portugal coincides with the cultural and educative reform taken by King João III (1521-1557). The strategic plan of Jesuit teaching, *Ratio Studiorum*,[2] was adjusted to suit the national circumstances and complied with the royal request to open the institution to the public. It provided lectures on the mathematics necessary for techniques related to the art of navigation. This departure from the pedagogical norm resulted in a tendency to emphasize the practical applications of science, avoiding higher levels of abstract theorization, and the use of vernacular instead of Latin in classes.

The organizational level of the network of colleges developed skills of communication and cooperation, gathering and exchanging scientific information, but was conditioned, however, by scholastic theology intended to glorify Catholic Church. Scientific contents and the aggregation of new arguments were established under the solid principles of the Counter-Reformation, where knowledge was managed in an agreement between faith and reason tending towards unity: Christ (verb) and God (truth) are converted into human speech and action. As such, the primary source of knowledge is, according to Aristotelian scholasticism, sensible reality, because the idea does not exist outside the world presented to the senses.

The *Ratio Studiorum* provided guidelines for mathematics education. The first concern was an introduction to Euclid's *Elements*, which was followed by an introduction to the parts of geography and astronomy that were related to geometry. Every month students were requested to solve a famous mathematical problem, which, in the presence of philosophy and theology students, was discussed in accordance with the ideological standards of the Jesuits, reflecting the approach to science as knowledge of the natural and recognition of the divine.

The restoration of Portuguese independence in 1640 led to a revision of the educational programs, redirecting them to suit the needs of the military war with Spain, and make up for the lack of specialized technicians. Subjects such as arithmetic, geometry and algebra became part of the programs providing the essential foundations for architecture (included in this institution by royal request), training architects to reshape the military infrastructure.[3]

With the resumption of international relations and positive economical development in the early eighteenth century, a profound change took place in the scientific scene in Portugal. Travel and scientific exchange were encouraged, as was the reform of educational institutions, where the Jesuits strengthened their influence and presence within the court, profiting from royal patronage. The *Aula da Sphera* played a central role in the Portuguese scientific educational panorama, which is testified to by the number of foreign teachers who held positions (about one-third of them all), making it possible to ensure that the curricula was continually updated.

The treatises by Inácio Vieira

Despite the cosmopolitanism and updated contents presented at the *Aula da Sphera*, Portuguese treatise output during the seventeenth century is limited and often devoid of the kind of reflection and scientific reasoning required to set forth subjects such as optics and perspective systematically.

With regards to perspective, related to the representation of reality and the instrumental processes of painting and architecture design, the "… texts didn't work as manuals where authors treat the practice of painting or drawing, but rather were conditioned by literary matters and exaltation of painting as a liberal art" [Melo 2002: 413]. According to Saldanha [1998: 85], the artistic treatise is deprived of the necessary scientific content by joining the apology in an affinity between literary and pictorial form.

In this situation, the theoretical reflection and systematization carried by Inácio Vieira is noteworthy. As teacher at *Aula de Sphera*, he drafted the *Tractado da Óptica* (1714) and the *Tractado de Prospectiva* (1716),[4] organizing the state of knowledge regarding the *nature and properties of vision*, the *errors and disappointments of view*, the *fundamentals of perspective* and *Reflection*. The treatises cover the geometric properties of visual rays, the principles of optical illusions and optical mathematical values inherent to the representation of space, along with the consequent practical applications in architecture and painting and the possibilities of projecting images in space.

Fig. 1. Inácio Vieira, *Tractado de Óptica,* 1714 (Biblioteca Nacional de Lisboa, Codex 5169), fol. 1

Fig. 2. Inácio Vieira, *Tractado de Prospectiva,* 1716 (Biblioteca Nacional de Lisboa, Codex 5169), fol. 1

Although he did not achieve an original scientific contribution, he formulated an operative step towards a full theoretical understanding of the subject, reiterating the positions of seventeenth-century Italian, French and German treatises: "…its omnivalent approach often introduces practical examples based on direct experience of authors of reference …" [Raggi 2004: 528]. These are the first Portuguese works to recognize the importance of proper representation of pictorial space, integrating *quadratura*[5] and scenography[6] design with current knowledge and practices, in accordance with the prevailing rhetoric and proselytizing mission of art in erudite Jesuit circles and European courts. Although the treatises remained in manuscript form, these syntheses were diffused in the college, through communication of its contents in the classroom, or by the circulation of manuscript copies made by students.[7]

Vieira's works condensed scientific knowledge developed over the 1600s, ranging from the dominance of theoretical and practical Italian experience (Pozzo) to the French essays (Dechales), German systematizations (Kircher, Schott and Scheiner) or pre-modern sources (Ptolemy, Euclid, Vitruvius, Alhazen, and Witelo). The two treatises discussed here developed a progressive investigation of the arguments relating vision and representation of a visible or imagined world, both dealing with the conditions of seeing, *óptica*, and the geometric interpretation of vision, *prospectiva*, simulating visual reality through its two-dimensional representation. They collect knowledge and tools converted into working logic for architects, painters and set designers, establishing the principles of a sensitive space.

Here, the exploration of perspective is affiliated with the effort of a *prós opsin euruthmia* (proportion in agreement with visual impression), gathering the beauty of form according to the subjective impression resulting from the deception of vision, once the ideal of absolute mathematical proportions is inaccessible to the architect by distortions triggered by the process of seeing. From this logical point of view, perspective does not refer to an exact image of the world, but to its appearance. Spaces of *quadratura* challenge perception, creating an intellectual awareness of a sensitive dimension of architectural work on which it is "built". Perspective raises a simulacrum structuring a seeming and mental reality.

Tractado da Óptica, *1714 (BN Codex 5169)*

The *Tractado da Óptica*, or Treatise on Optics, organizes the subject into two branches: the first physical, relating to the "*organ, or instrument of vision,*" and the second optical (or perspective), relating to "*visual things*". This identification structures the document. While physics treated the line "*as a physical object*", generally regarding anatomy and a physical understanding of vision and the viewed image, perspective treated it as a mathematical object, regarding an abstract system, converting how we see it into geometric-mathematical facts.

As such, the first part, "*the properties of the eye fundamental to optics*" (fols. 2 to 95), explores the anatomy of the eye; the second part, "*the nature and properties of seeing*" (fols. 96 to 246), examines the conditions of sight towards perception of displayed reality, presenting some solutions to rectify the perceived form. The end of this second part, an Appendix entitled "*From some propositions pertaining to this matter*" (fol. 196), is based upon the understanding of differences between real and perceived image as result of visual perception, analyzing a possible manipulation of perceived image and reality. Material collected here constitutes the support for the speculations of the third part, "*Of the mistakes and disappointments of view*" (fols. 247 to 375), which explores distortion of reality for the purpose of deception.

It was through the exploitation of optical illusions that the Church enthusiastically promoted the development of mechanisms to support Counter-Reformist rhetoric. In "*The precious appearances made by nature*" (fol. 320), natural phenomena are interpreted, for subsequent application in "*How to represent by art what nature displays*" (fol. 329), developing a "Handbook of miracles," turning knowledge into a capacity for the deception of vision and obstructing viewer's reason "*without suspicion of diabolical art*".

At the beginning of the third part, Vieira seeks to understand how our senses condition our relationship with bodies displaced in space. How does the deception of our senses interfere with judgment of reality?

> ... *by corrupting the senses, judgment, depending too much on them, would fall into error ... as our reasoning was in many things dependent on senses, if they do not present the object's truth, reasoning can't be undeceived* [Vieira 1714: 247-248].

The mistake of judgment regarding reality is a consequence of the deception of senses conditioning reasoning and understanding of the world. Thus, Vieira refers to painting as a condition of deception and simulacrum of the act of seeing; he maintains, however, the differences between representation and the image "painted" on the back of the eyeball, as a process of seeing.

Philosophical discussion in the Baroque age compared the real world and the perceived world in which visual illusions offered by perspective science as playful instruments became overburdened with religious and scientific values, establishing a feeling of uncertainty. This uncertainty over sensitive reality was criticized by Descartes in *Discours du méthode* (1637), putting senses under the doubt of reason. In his discourse, the understanding of the world, based on a universal mathematical basis, order and measure, should ignore everything that comes from senses and tradition. This rationalization in the search for truth was developed in contrast to the Aristotelian empiricism, and especially to the overestimation of the five senses, especially vision.

In Inácio Vieira's treatise, reflection is built upon an exaggerated value of vision, considering the mistakes of the eye as mistakes of reason, once rational judgment, according to the Aristotelian conception of natural knowledge, is blocked from a direct relationship with the outside world depending on the senses. Furthermore, the excessive rationalization of Descartes is downplayed by Vieira, given his agreement with the Jesuit concept of empirical science: "*This very same issue is deeply developed by Descartes, but since he has never exercised it, it misses many things*" [Vieira 1717: 690].

The treatise goes into issues of visual angles when, without referring directly to Euclid's propositions, it explains them through the observance of reality. The section entitled "*From some curious problems for deceiving and undeceiving sight*" (fol. 263) introduces practical arrangements of "error". From this point forward, the author begins a long statement of optical/perspective problems leading to the production of anamorphosis on a wide range of surfaces, with different purposes, along with several procedures for their resolution.

In "*How to draw some square images*" – the first problem (fol. 264) – the images produced to entertain senses are taken from Dechales' *Opticae, libro secundus*.[8] The second problem (fol. 267) further examines the construction of anamorphic figures by the "*outline of a deformed image, which viewed from a particular place seems perfect and well developed*"; the text coincides with the statement of Dechales and the drawings are similar to table 12 of Niceron's treatise.[9]

Fig. 3. Inácio Vieira, *Tractado de Óptica*   Fig. 4. Inácio Vieira, *Tractado de Óptica*
(1714), fol. 247   (1714), fol. 275

"*Deformation on the conic from the images in the flat plane*" (fol. 279) guides the reader on how to produce a perspective device, automating perspective through the materialization of its geometric elements (*board, point of view, visual ray*), without the use of geometric-mathematical procedures. Kircher's *mezoptico*[10] instrument (explained via Schott) has a dual function, making it possible to draw what is visible, the space beyond the framework and to project image into space through "*light and shadow*" or "rope." By placing a light source at the viewing point, the image projected into space is admitted in any surface and with any level of deformation without using the abstract procedures based on the distortion of grids presented in the previous chapter. Later in the treatise, in "*How to project any figure in interrupted planes to be seen from a certain place*" (fol. 317), the discussion returns to perspective machines, with references to Dürer and Maignan (also quoted via Schott), and including their expedients for anamorphic construction.

Concluding the variety of possible implementation for anamorphic constructions, the twelfth chapter, "*Deformation of images in any other planes mainly in the interrupted*" (fol. 314) explores the deformation of an image on a specific space, entering into the explanation and practice of *quadratura*:

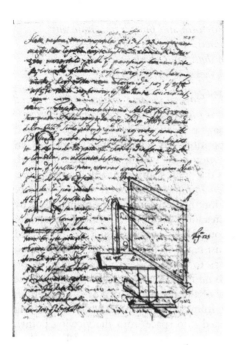

Fig. 5. Inácio Vieira, *Tractado de Óptica* (1714), fol. 280

Fig. 6. Inácio Vieira, *Tractado de Óptica* (1714), fol. 283

Fig. 7. Inácio Vieira, *Tractado de Óptica* (1714), fol. 306

From what we have said, it is possible to deform and project images in any type of plane and observe similar geometrical deformations. However, mechanically it can be worked easily by light and shadow, by string, or with optical beams directed to the plane [Vieira 1714, 314-315].

The first problem of this chapter (fol. 315), "*How to deform images in interrupted planes which by approaching or moving appear as projected parts, and are seen perfectly only from one place*", provides a method to image projection into space to be observed (according to the prototype) from a specific position. To achieve this, Vieira uses an optical instrument of Kircher which Vieira calls *mezoptico*, which, set in place and with strings extending from the optical spot, outlines the desired image on any kind of plane, "*... there is no surface, no matter how misshapen and discontinued, on which it can't be done*" [Vieira 1714: 317]. Here a second use for the *mezoptico* instrument arises, by reversing the operation of Dürer's perspective machine. While in a previous paragraph, projection was based on image delineation through light beams passing the perforated points in an opaque support; here, despite the insistence on light spot and the projection of points, the author refers to the use of strings to determine each point in space on surfaces that receive the figure. In the *Tractado de Prospectiva* we will again find these two steps. However, these procedures grew out of the misgivings of Pozzo between an ideal system (based on projecting the image from a light source) and the existing circumstances (construction from strings stretched into space from the viewing point), consequent to technical constraints and space dimensions in which image is delineated.

The application of these principles makes it possible to achieve a high degree of illusion when, for example, "*... a beam can be painted so that it really seems to be set on the walls ...*" as in the case of "*... the entrance hall of S. Vicente monastery, the choir of S. Francisco and the tomb of the Cathedral ...*", referring to the existing examples of *quadratura* in Lisbon:

> *We may notice that paintings executed in this form have a special beauty and marvel. But, once the observer is displaced from the viewing point, painted constructions appear broken, smashed and totally gone, columns and porticos seem to move and fall; although, returning the observer to this spot, image become into its perfect figure, similar to its prototype* [Vieira 1714: 317].

Here, Vieira shows, according to the construction from a single *viewing point*, its illusory effects and how the image changes according to the observer's movement. Represented architectures arise with the *body*, as a tectonic fact, and are disrupted or collapse as the observer moves. This fact leads us to the ideological values exposed by Pozzo in the dogmatic defence of a unique point of view. According to Pozzo, contemplation of transcendent facts of faith would only be perceived when the observer was placed in the right location – the path of faith – outside of which the sublime crumbled. This thought integrates the illusion of a sublime truth in the same sense of contemplation expressed by Ignatius of Loyola in his *Esercitia Spiritualia*: "For, if the person who is making the Contemplation, takes the true groundwork of the narrative, ... he will get more spiritual relish and fruit ..." [Loyola 1538: Second Annotation].

Tractado de Prospectiva, *1716 (BN Codex 5170)*

In the Prologue (fol. 1)of the *Tractado de Prospectiva*, or Treatise on Perspective, the author identifies the scientific domain of perspective as the field of mathematics related to everything that concerns the eye. Its use is restricted to the representation of objects that:

> … *drawn this way will form into the eyes a very similar image of the object, the very same that the object would form if presented to the eyes; from which arises, that all painting belongs to perspective* [Vieira 1716: 1].

Organized in six Tables, the treatise systematizes arguments related to perspective science, from ground rules to complex practical application: "*Fundamentals of perspective*" (fol. 4); "*Projective scenography*" (fol. 34); "*Appearances of bodies in any sort of position*" (fol. 222); "*Ceilings, and vaults*" (fol. 270); "*Composition of various boards, reflection and shadows*" (fol. 298); "*An instrument useful in practice*" (fol. 333). This sequence is interrupted by an "*Opportune tour in civil architecture, single line about the orders from this Science*" (fol. 90), which explores architectural orders, principles of composition and perspective interference into architecture. At the end of the treatise, a supplement is added to explain the "*Method of Brother Pozzo dealing with spiral columns*" (fol. 360).

In the first table, "*Fundamentals of perspective*", the relationship between the observed reality and the represented one is explored, with definitions of basic concepts (*Picture plane, Base line, Main convergence point, Horizontal Line, Distance line, Distance point, Mainline, Ray, Objective Line, Appearance of objective line, Objective plane*), together with its main theorems. The discussion is practical rather than speculative, since demonstration of fundamental concepts is based upon experience with the perspective device (which was previously presented in the treatise on optics).

The second table, "*Projective scenography*", is about the practice of perspective representation: achievement of plan perspective and, by obtaining heights, definition of construction volume.

While the first table followed the structure of Dechales' treatise on perspective, the second table abandons this for cross-reference, comparing different procedures for each construction. It begins by giving guidelines for establishing the observer's position in front of the object. "*The location of the observer must be outlined, viewing point, that according to Andrea Pozzo is the only one … which easily allows us to obtain the main convergence point and distance points*" [Vieira 1716: 36]. Pozzo upholds the uniqueness of the point of view[11] and, through a triple space projection (plane, section and elevation), he sets out the fundamentals of perspective for the setting up of illusory images in a real space.

The presence of Pozzo's procedures in Inácio Vieira's treatise, through long quotations and explanatory figures, is important in the Portuguese panorama of perspective science history, confirming just how up to date the work was, since the first Portuguese translations of Pozzo's *Perspectiva Pictorum et Architectorum* (1693) only appeared in the 1730s. The inclusion of this source is an acknowledgement of Pozzo's practical procedures and its dissemination through Vieira's educational efforts at *Aula da Sphera*.

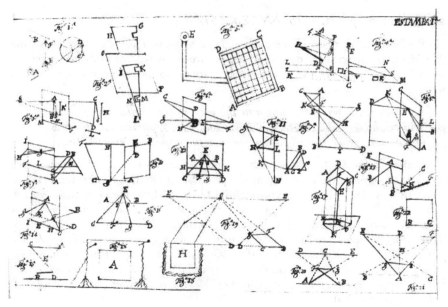

Fig. 8. Inácio Vieira, *Tractado de Prospetiva* (1716), Table 1

The practice of perspective begins throughout plane figures on the ground floor facing several parallel constructions: *"This practice is from Andrea Pozzo ... Father Dechales tell us to do this way"* [Vieira 1716: 36]. While in Pozzo's example, the square is obtained by determining the depth of an edge; according to Dechales, the square is constructed by the perspective of its diagonals. Both cases are translations where nomenclatures and constructive sequence coincide integrally with the sources.[12]

In the sequence of obtaining grids and pavements in perspective, Vieira interrupts the explanation of this argument to engage an *"Opportune tour in civil architecture"*. He deviates from his main sources justifying that perspective

> ... *can't take a step without knowing the proportional principles and measures of the five architectural orders ... perspective must be taken along with architectonic knowledge, ... once perspective is demanded to outline in plane, with colours and brush, the gallantry of Architecture ... that the well-tempered architect has put into solid ...* [Vieira 1716: 90-91].

It explains the need for architectural representation to obey the same compositional principles as *"solid"* architecture. In this *tour*, which quotes Vitruvius, Serlio, Vignola, Palladio, Scamozzi and Dechales, the sequence and content of the definitions presented lead us to believe that Vieira knew Ferdinando Galli Bibiena's *L'Architettura Civile* of 1711. If this source could be proved, given the already proven inclusion of Pozzo's procedures, the up-to-date nature of the treatise would be truly remarkable.

In his architectural treatise, *Regola delli cinque ordini d'architettura* (1562), Vignola announced the preparation of a treatise in perspective, *Le due regole della prospettiva pratica* (1583), which was to make it possible to use perspective to evaluate optical distortions in constructions, once the rules for architectural ornaments were established. The comprehension of optical distortion brings up the relation between reality and appearance identified by Plato, in *The Sophist*, in which image production lies in the fact that to show an object one has to show it as *it is not* in reality. Plato compares two categories of imitation: one that reproduces reality in its true metric proportions, producing awkward images due to optical distortion; the other in which artists neglect true proportional ratios, at the expense of material truth, given precedence to appearance. It follows that representation of reality re-creates the appearance of things; just as in architecture the correct appearance of the proportion of elements is only possible through mechanisms of perspective.

In the fourth table, *ceilings and vaults*, the author explores questions about the practice of *quadratura* and the representation of illusory architectures leading towards the formulation of a sensitive space. *"With such a painting, we fool the eye, and so we present the rudiments required to outline it in roofs, rooms, vaults or arched surfaces"* [Vieira 1716: 270]. In essence, this table follows the *Liber Quintus* of Dechales' *Perspectivae*, introducing cases concerning the Portuguese practice *of quadratura*, circles of influence and authors.

Vieira exposes the *universal principle for delineation of any appearances in ceilings* (fol. 281) operating the passage from *vertical painting to horizontal* (fol. 281), to which is applied *"... a repetition of the already taken precepts... from which all appearances raised upon any sort of roof will be painted following the same method applied to vertical surfaces ..."* [1716: 281]. The difference lies in the nature of the picture plane that can be flat, *"... circular, elliptical, or often consists of arranged planes and curves, forming an irregular surface on which it is very difficult and often impossible to find a bottom line, an horizontal line, and distance points, according to the universal practice..."* [1716: 283].

The *universal practice* mentioned consists of the image projection by means of grids laid over the architectural surface, allowing the transference of a delineated perspective. But, in Vieira's illustrations (as in Dechales' treatise), the deformation of the grid doesn't appear to be constrained to a specific point of view; it gives the impression of an orthogonal projection rather than a conical projection of the grid, as explained in the text. Vieira's manuscript doesn't address these inconsistencies between the explanation and the illustrations (figs. 290-291 of the manuscript), faithfully following Dechales' procedures for projecting the grid according to a specific point of view.

Exploring the construction of an underlying grid, Vieira proposes three possibilities for its projection onto irregular surfaces. *"For this, strings may be used... or by sight and nothing else, or at night by putting some light in the viewing point..."* [Vieira 1716: 283-284]. Besides construction through stretched strings or light rays, the author introduces a third hypothesis, *"by sight and nothing else"*, referring to the accomplishment of perspective through trial and error based upon practical experience, while acknowledging that geometric-mathematical procedures would be implicit.

Through deformation of the underlying grid, it is possible to transfer *"... painted objects from each plain square (drawn in the prototype) to the corresponding projected quadrilateral, and we'll obtain the overall finished work"* [Vieira 1716: 284]. The grid's transposition from a drawn prototype to an architectural surface is explored by Vieira according to the vault's configuration and the composition of imaginary architectures, leading us to a new hypothesis on speculation over practical procedures of *quadratura*.

Fol. 292 shows a curved portion of a vault, ABCD, onto which it is intended to transfer an image previously drawn on a flat panel, CDFG, to be seen from the viewpoint E. According to the author, the use of light rays only occurs *"To better understand what is said..."* [Vieira 1716: 284]. They are used to demonstrate the grid's geometric transformations when it is projected from a flat surface onto a curved one. Simultaneously, the reference to the stretching of ropes increases the possibility of understanding geometric procedures through the materialization of abstract elements: *"... instead of the lines stretch out cords..."* [Vieira 1716: 284]. Meanwhile, Vieira identifies the basic geometric elements and their function: E - point of view, J - placed upright over E, which serves to divide the curved surface portion, ABCD, into quadrilaterals corresponding to the flat board, CDFG (figs. 9-10).

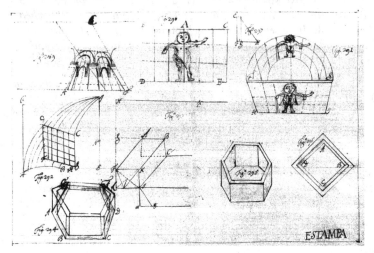

Fig. 9. Inácio Vieira, *Tractado de Prospetiva* (1716)

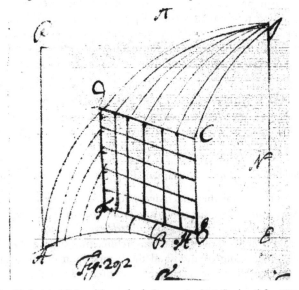

Fig. 10. Inácio Vieira, *Tractado de Prospetiva* (1716), detail fig. 292

The main problem of drawing architectural perspectives on curved surfaces concerns the transfer of grids, so that *"... irregular quadrilaterals formed on the architectural surface may appear to the eye equal to the squares of the panel CDFG"* [Vieira 1716: 284]. As such, this transposition and consequent deformation of the grid is organized in two steps concerning vertical and horizontal lines.

To project vertical lines CG and DF the author refers to Proposition 18 of Book 11 of Euclid's *Elements* [Vieira 1716: 285] (If a straight line be at right angles to any plane, all the planes through it will also be at right angles to the same plane [Euclid 1956: vol. III, 302]). Applying the Euclidian proposition, Vieira demonstrates that if CG and DF are vertical parallel lines to EJ, planes projecting these «shadow» lines, and containing EJ, are vertical surfaces: *"... where, by any of the identified lines and the sight E, an imagined plane will necessarily be vertical, through which shadow of the same line will be found..."* [Vieira 1716: 285]. As such, projection of the grid's vertical lines over the vault surface is determined by the vertical planes containing EJ as the projection of all intermediate parallel lines: *"...All vertical planes passing through the point of view E have the common section line EJ: so do all the shadows of the lines DF, FC, and the other parallel to this common point J"* [Vieira 1716: 285].

From this explanation, it may be inferred that it is possible that the projection of vertical lines was made either by calculation of the planes or by practical determination through triangulation by strings. It follows, at least hypothetically, that it is possible to drawn lines projection on the dome using only the space available between the scaffolds and the vault, without necessitating the unwieldy stretching of strings from the actual point of view located well below the surface on which spatial illusion was to be carried out. This hypothesis could pave the way for a consideration of an alternative practical procedure for the construction of an architectural perspective,[13] at least regarding semi-cylindrical or compound vaults resulting from the intersection of four semi-cylindrical with a flat central area.

The author goes on to say:

> There is no difficulty, given three points of the same plane, EDJ, to extend the first string ED and, from the point J, extend another string that touches the first string at any given point so it will produce in the vault a point belonging to such plane – with this method we'll have the grid vertical lines [Vieira 1716: 285].

As is usual with Vieira, the discussion starts with general rules and goes towards practical implementation, providing information for the accomplishment of perspective.

Regarding the projection of the grid's horizontal lines, Vieira says that once horizontal lines are equally spaced in the prototype, it is possible that to divide vertical line FD, used to obtain the grid's verticals lines, into equal parts (e.g., by knots tied on the rope itself) to obtain horizontal spacing: *"... a rope can be suspended from point D with its dividing knots, and we will have a quadrilateral corresponding with the prototype's squares"* [Vieira 1716: 285]. He continues the explanation stating that:

> ... each of the quadrilaterals obtained is the shadow of the matching square. So, they will arrive to the sight E by the same rays as the squares of

the prototype: so painted objects in such quadrilaterals result in the same vision as in the prototype. The same method will be used when delineating an object on a vertical wall, but meant to be viewed obliquely in such a way that you cannot find the main point in it [Vieira 1716: 285-286].

The procedure set out by Vieira leads us, in the context of the practice of *quadratura* in Portugal, to the type of illusory architectural composition and the vaults on which *quadratura* is applied. The architectural composition is carried along the surface of the vault, above the surrounding moulding of the actual spatial box, until it reaches a constant height around a virtual perimeter. The level of that new virtual moulding may match the back of a vertical grid in DC, according to Vieira's scheme, leaving the vault's central section available for the depiction of an open sky, or providing space to install the *quadro riportato*. The grid's deformation, according to this assumption, can lead to the perspective discrepancies found in some *quadratura* paintings, through the use of "*sight and nothing else*". Moreover, with reference to the structure of real and illusory space in Lisbon's Menino Deus church (ca. 1730), might the four focal points (one for each side of the main composition) detected in previous studies, be a consequence of this procedure?

Let us now leave the practical implementation of these procedures and move forward in Vieira's perspective treatise.

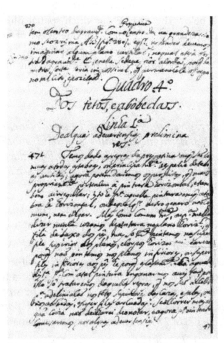

Fig. 11. Inácio Vieira, *Tractado de Prospetiva* (1716), p. 270

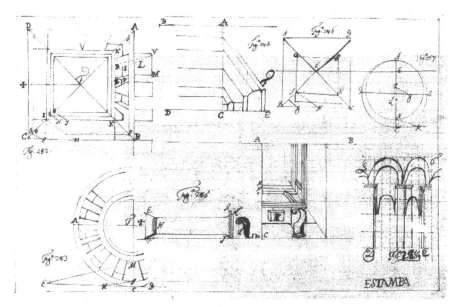

Fig. 12. Inácio Vieira, *Tractado de Prospetiva* (1716) no number

Spatial correction is introduced with "*The amendment of the bodies*" (fol. 289), in which the author addresses the "*Amendment of buildings*" (fol. 289), "*Amendment of the five-sided room*" (fol. 291), "*Amendment of the room with a inclined wall*" (fol. 292), "*Amendment of the very low ceiling*" (fol. 292), "*Amendment of sloped ceilings*" (fol. 293) and "*How to make a room, or porch seem bigger and outline a entire building, from which all the parts are seen*" (fol. 294). Vieira informs the audience at *Aula da Sphera* about how perspective can be orchestrated for spatial conceptions and the correction of architectural forms that interfere with the perception of built space. The process is essentially the application of an imaginary picture onto surfaces of the space so that, from a certain viewpoint, the desired simulation can be observed:

> *Every time a ceiling is very low and we want it, from a certain point, to appear higher, we must delineate a higher ceiling and continue its walls with false windows ... When ceilings are sloped, it follows that one of the walls is higher than the other and the side walls will gradually deviate from equality. To amend this defect, the missing parts of the walls that lack equality must be outlined on the sloping roof ...* [Vieira 1716: 292 - 293].

This application leads us towards the correction of physical space and the dramatization of reality based upon the illusion of the senses and therefore of reason. After defining perspective drawing, the author adds a deep appreciation of fostering depth based upon line and colour definition, amplifying illusion according to the precepts of atmospheric perspective. In this case, Vieira refers to his own sensitive experience in front of Bacharelli's work in Lisbon, "*... after I saw that feast of sunlight in St. Vicente's entrance hall quadratura painting, it doesn't seem so impossible for anyone possessing good art*" [Vieira 1716: 294].

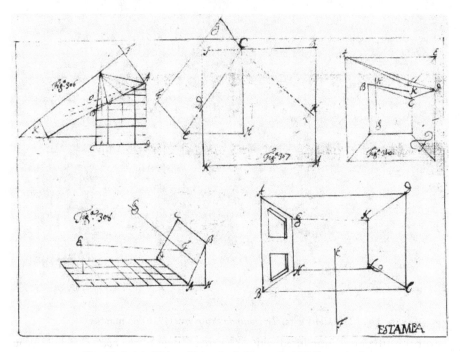

Fig. 13. Inácio Vieira, *Tractado de Prospetiva* (1716) no number

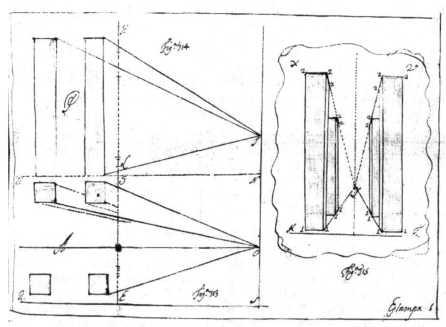

Fig. 14. Inácio Vieira, *Tractado de Prospetiva* (1716) no number

In the fifth table, "*Composition of various boards, reflection and shadows*", Vieira presents "*some more praxis from Brother Pozzo*" (fol. 296), setting forth other applications of perspective as tools to achieve a sensitive space: ephemeral devices and scenography. Pozzo's procedures are presented as resulting from his personal practice on painting, architecture and scenography, rather than abstract speculation, associating his construction process to the *costruzione legittima*. However, this system is more complex than the previously presented synthesis based on convergence and distance points. The same process of exchange occurs between Pozzo's Book I and II, where drawing is controlled with the aid of strings to determine the visual cone section. "*Instead of drawing lines with a pencil … we should apply to the view a thin string, or a drawing scale*" [Vieira 1716: 303]. This drawing process brings us to the perspective practical technique applied in image projection onto an architectural surface: ropes are extended from a specific point to the surface, transferring the desired prototype (fig. 14).

At the end of the treatise Vieira presents the pantograph, "*a useful tool for practice*" (fol. 333), explained in three chapters: "*Construction of the instrument to delineate*" (fol. 333); "*Perfect conformation of parts of the aforesaid instrument*" (fol. 340) and "*Use of this instrument*" (fol. 354). Christoph Scheiner's work, *Pantographice seu ars delineandi* (1631), is quoted at length, and Vieira praises this instrument whose function is "*… to design and launch on any surface any object's image… with infallible art …*" [Vieira 1716: 357]. From this statement we are obliged to ask: Can the pantograph serve *quadratura*? Can the instrument be used to transpose images from a prototype onto full-scale panels?

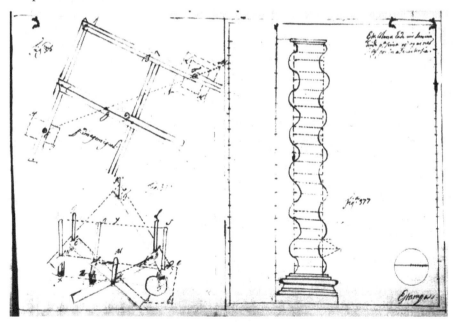

Fig. 15. Inácio Vieira, *Tractado de Prospetiva* (1716) no number

To the pantograph and the *mezoptico* instrument, we must add another projective device: the magic lantern described in Vieira's treatise on catoptrics (1717). There, in "*Appendix 1st*" (fol. 681), "*The Magic Lantern*" (fol. 681), "*From air, sun, and other appearances*" (fol. 685) and "*Method for making glass*" (fol. 689), the device displayed in

Lyon by a Danish scholar and seen by Dechales is revealed. According to the source, Vieira describes the relationship observed between the scale of the projected image and the distance to the wall, or screen, which may be adjusted so that "*... the height of the image equals the size of a man*" [Vieira 1717: 682].

Contrary to what is found in Kircher's and Dechales' iconography of the magic lantern, Vieira does not suggest a specific application of the apparatus, leaving this to his readers: "*From observation of these instruments a universal manufacture and use are taken ... and as I write to ingenious masters, they alone can work out the truth of this manufacture...*" [Vieira 1717: 683]. However, he admits that all the arguments relative to this *machina* derive from the same principles set forth above in optics, perspective and catoptrics.

The visible world and construction of an illusory reality

Inácio Vieira's treatises evidence an extended reflection on optics and perspective to support image distortion and answer the formulation of a sensitive space, imaginary architectures at the service of the triumphant Church and Counter-Reform rhetoric. As Vieira writes, in his treatise on catoptrics, "*God our lord laid our intentions to his own greater honour and glory*" [Vieira 1717: 2].

During the seventeenth and eighteenth centuries, discussions regarding visualization and appearance were situated in the context of contributions from optics and perspective along with political and religious constraints that govern Counter-Reform imagery and spatial frameworks. As Kemp mentions, "The demands of the catholic reformers brought a renewed insistence on theological ends over and above the artistic means" [Kemp 1990: 85]. Theological boundaries were imposed on artistic and scientific issues, inducing a clear confrontation between Idea and Science: the idea of post-Tridentine doctrine based in Aristotelian space versus the scientific approach to a uniform and infinite space.

These contradictions are at the origin of a conflict between a symbolic vision and a mechanic vision of the world (resulting from Cartesian modernity and the scientific revolution). This conflict legitimizes the emergence of perspective understood simultaneously as a cognitive model, validating a scientific representation of the world, and as a symbolic configuration maintaining the Aristotelian approach of senses over knowledge: "During the seventeenth century, the space occupied by man was not homogenized, and the primacy of perception as the foundation of truth was hardly affected by the implications of this new science and philosophy" [Perez-Gomez and Pelletier 1992: 28].

Compared to the Renaissance, the Baroque provides the transition from symbol to allegory [Argan 1989: 7]. As such, the seventeenth century faces the passage from "imagined" – the Renaissance conception of an ideal world – to "imagination" – the vision of the world as an entity floating between reality and wonder. The symbol, mark of cognitive rationalization based on code and inducing significance of the Renaissance sciences, is replaced by the allegory, which establishes ambiguous references incorporated into Counter-Reform culture, placing the subject between reality and the sublime.

Baroque culture lies in infinite possibilities of connection between spirit and science. Spatial design in this period magnifies new aesthetic criteria integrating in constructive material a synthesis between science and theology: the mathematical-geometric abstractions of the design with the cosmic conception of a triumphant Church.

Within the spirit of these paradoxes is *quadratura,* transposing architecture representation from a two-dimensional picture into a three-dimensional simulacrum, together with metric spatial properties, and contributing to a transmutation of its perception. The space represented in *quadratura* is freed from the constraints of physical construction, and, in the fusion between reality and appearance, sets up a single entity globalizing physical and represented construction into a new appearance offered to perception. Space is replaced by its image, decomposed and recomposed in a dreamlike atmosphere where architecture, real and represented, appears as a spectrum mesh of lines and colours to be lost again in that whirlwind.

As a space and cosmos allegory, *quadratura* constitutes an extension beyond the data provided by building measure. This presents us with a space based on sensory experience triggering an experience of architecture that is perceptive rather than rational. With the achievement of imagination into visual image, the imaginary becomes part of reality. According to the embodiment of illusory images, it can be declared "… that to perceive something is not only to register it mentally, but to be solicited by it; the mind must create new systems of reference adapted to the perception of objects which are no longer 'natural', but artificial products of man" [Argan 1989: 55].

As such, the production of the Baroque image is not centred on the object but extended beyond it. As far as the architectural space is concerned, perception is led from the building's spatial constraints towards an image of unlimited space. The grandeur and monumentality of architecture isn't just a result of built form, but of simulated architecture, where *quadratura* generates a second, unreal nature, which in its likeness is indistinguishable from reality. Space, more than limited, is considered elastic as the perception of simulated architecture, referring the observer to a space whose proportions are variable, expanding it and taking away its uniformity.

Conclusion

Portuguese architectural production in the seventeenth and eighteenth centuries underwent a renewal through the assimilation of international models, including a new feature of spatial research: *quadratura* painting. The *contrappunto* of architecture/*quadratura* is intended as an action in which real (constructed) and illusory (represented) space intertwine, creating a complex reality.

The simulated of *quadratura* space constitutes a challenge to perception, generating new dimensions in architecture. This establishment of a new vision of the world results from the interaction of mathematics, optics and geometry, which can't be reduced to the construction of a physical and tectonic reality, but must include an apparent reality based upon the power of the projective image. The polyphonic nature of the Baroque is determined not only by the built formal exuberance but also by the participation of different arts and sciences in the conception of a global space.

In Portugal, the operative enunciation towards a full theoretical understanding of the subject applied to the configuration of a sensitive space taken by Inácio Vieira constitutes a seed that, along with the work of João Antunes (a definitive reference of architectural renovation, introducing models from international baroque adjusted to a national constructive tradition), simultaneously with the practical essays of the Florentine quadraturist Bacharelli in Lisbon (introducing and updating Baroque *quadratura* painting) paved the way for new developments in space conception.

In this context, the architectural project, along with *quadratura,* questions new dimensions on Portuguese Baroque space. The plurality of visual centres involves the viewer in a fluid space where seductive potential of the Baroque is accentuated by the integration of optical phenomena guiding space towards a new sensitive and illusory experience.

Notes

1. The term *quadratura* was established in the sixteenth century to describe the development of pictorial representation of illusory architectural environments, especially in illusionistic ceiling painting. It exploits knowledge of perspective and optics in order to deceive the eye, simulating a sense of depth and space.
2. The *Ratio Studiorum* was a regulatory document addressed to Jesuit teachers, promulgated in 1599, concerning the nature, extension and obligations over the curricula in order to unify pedagogical procedures among colleges of the Company.
3. Subjects ranged from astronomy to cosmography, geometry based on the study of Euclid's *Elements*, to arithmetic, algebra, plane and spherical trigonometry, issues applied to navigation, hydrography and cartography, optics, perspective and scenography, gnomonics, statics and hydrostatics, architecture and military engineering and other related topics, such as pyrotechnics and ballistics, etc.
4. Vieira's other treatises include Chiromancer – 17?? (BN Cod 7782); Astrology (ANTT ML 2122) and Chiromancy – 1712 (BN Cod 4324), both collected in M.L. 2132 from the Arquivo Nacional da Torre do Tombo (ANTT); Astronomy – 1709 (BN Cod. 2111), Astronomy – 1710 (ANTT ML 2044); Mathematical Pyrotechnics – 1705 (BN MSS. 22), Catoptrics – 1716 (BN Cod 5165/1); Dioptric – 1717 (BN Codex 5165/2); and Hydrographical or art of sailing – 1712? (BN Cod 5171), and also orientation of thesis: *Perspectiva Mathematica*, by José Sanches da Silva (1716) and *Conclusoens mathematicas de huma, e outra esfera e Architectura Militar Munitoria, e expugnatoria*, by António Gomes de Faro (1710).
5. In spite of some sixteenth and seventeenth century instances, one can say that Baroque *quadratura* was introduced in Portugal with the work of Vincenzo Bacharelli (1672-1739). This Florentine painter stayed in Lisbon from 1702 to 1719 and brought the updated model of the Bolognese School to Portugal. He painting the ceiling of the lobby in the St. Vicente monastery during this period (1710).
6. The dynamics of *Aula da Sphera* is revealed, in addition to teaching and scientific publication, by the organization of activities (the creation of ephemeral apparatuses for religious festivals or theatrical performances) that enables practical implementation of taught theories.
7. In this regard other manuscripts (BN codex 1869, codex 2127, Codex 2258, Codex 4246) must be taken into account. These are compilations of notes, which circulated among students and proved to be an important source of transmitted contents.
8. *Tractatus XVIII; Opticae – Liber Secundus; PROPOSITIO LXIX – Tesselatus, imagines consirvere*. The explored principle coincides with paragraph 94 of Vignola/Danti treatise, "*Come si faccino quelle pitture, che dall'occhio non possono esser viste se non riflesse allo specchio*".
9. *Tractatus XVIII; Opticae – Liber Secundus; PROPOSITO LXX – In plana superfície, imaginem difformem d'lineares qua ex certo & determinato loco, omnibus futs partibus absoluta videatrur.*
10. *Mezoptico* was the term used by Inácio Vieira to describe the device for drawing and projective designed by Athanasius Kircher.
11. As expressed at *Figura Prima* of book I of Pozzo's treatise, "*Explicatio linearum Plani & Horizontis, ac Punctorum Oculi & Distantiae*".
12. Pozzo, *Figura Secunda – Modus delineandi optice Quadratum*; Dechales, *Proposito I – Quadratum directe oppositum describere.*
13. This is an alternative procedure to the hypotheses presented by Daniele Di Marzio regarding the architectural perspective in the curved vault of Sala Clementina in the Vatican palace [Di Marzio 1999: 163-166].

References

ALBUQUERQUE, Luís. 1973. *Para a história da ciência em Portugal.* Lisbon: Livros Horizonte.

LOYOLA, IGNACIO DE. 1538. *Esercitia Spiritualia.* Rome. Engl trans: *The Spiritual Exercises of St. Ignatius of Loyola,* Father Elder Mullan, S.J., trans. New York: P. J. Kenedy & Sons, 1914.

ARGAN, Giulio Carlo. 1989. *The Baroque Age.* Geneva: Skira.

BROOKE, John Hedley. 2003. *Ciência e Religião. Algumas perspectivas históricas.* Porto: Porto Editora.

CAETANO, Joaquim Oliveira and SOROMENHO, Miguel. 2001. *A ciência do desenho. A ilustração na colecção de Códices da Biblioteca Nacional.* Lisbon: Biblioteca Nacional.

DI MARZIO, Daniele. 1999. La Sala Clementina in Vaticano. Procedimento per la costruzione diretta della prospettiva su superficie curve: ipotesi teorica e verifica sperimentale. Pp. 152-177 in *La Costruzione dell'Architettura Illusoria,* Riccardo Migliari, ed. Rome: Gangemi Editore.

EUCLID. 1956. *The Thirteen Books of the Elements.* 3 vols. Thomas Heath, trans. New York: Dover.

FARRELL, Allan. 1970. The Jesuit Ratio Studiorum of 1599. Presented at the Conference of major superiors of Jesuits, Washington, United States, July 21, 1970.

KEMP, Martin. 1990. *The Science of Art, Optical themes in western art from Brunelleschi to Seurat.* New Haven and London: Yale University Press.

LEITÃO, Henrique. 2008. *Sphaera mundi: a ciência na aula da esfera: manuscritos científicos do Colégio de Santo Antão nas colecções da BNP.* Lisbon: Biblioteca Nacional.

———. 2002. "Jesuit Mathematical Practice in Portugal, 1540-1759". In *The New Science and Jesuit Science: Seventeenth Century Perspectives,* edited by Mordechai Feingold, 229-247. New York: Springer.

MELO, Magno Morais. 2002. Perspectiva pictorum: as arquitecturas ilusórias nos tectos pintados em Portugal no século XVIII. Ph.D. thesis, Universidade Nova de Lisboa.

MURTINHO, Vítor. 2000. *Perspectivas: o espelho maior ou o espaço do espanto.* Coimbra: Departamento de Arquitectura da FCTUC.

PEREZ-GOMEZ, Alberto and PELLETIER, Louise. 1992. Architectural Representation beyond Perspectivism. *Perspecta* 27: 21-39.

RAGGI, Giuseppina. 2004. Arquitecturas do engano: a longa construção da ilusão. Ph.D. thesis, Universidade de Lisboa.

SALDANHA, Nuno. 1998. A muda poesia. As poéticas da pintura no Portugal de Seiscentos. In *Bento Coelho 1620-1708 e a Cultura do seu Tempo,* Luís de Moura Sobral, ed. Lisbon: IPPAR.

VIEIRA, Inácio. 1714. *Tractado de Óptica.* Codex 5169, Biblioteca Nacional de Lisboa.

———. 1716. *Tractado de Prospectiva.* Codex 5170, Biblioteca Nacional de Lisboa.

———. 1717. *Tractado da Catóptrica.* Codex 5165, Biblioteca Nacional de Lisboa.

About the author

João Cabeleira is an architect and geometry teacher. He earned a degree in architecture from the Faculty of Architecture of the University of Porto (FAUP) in 2002. In 2006 he completed a Master's degree in Methodologies of Intervention in the Architectural Heritage at FAUP, with a thesis entitled "Versatility and changing at public space in historic city. Methods of intervention". He is currently preparing his Ph.D at the University of Minho (EAUM). The research is entitled "Imaginary Architecture: Real and illusory space in Portuguese Baroque", with advisors João Pedro Xavier (FAUP) and Jorge Correia (EAUM), and examines architecture and perspective treatises in order to identify intersections between science and design processes for imaginary architectures. Licensed as an architect at the Portuguese College of Architects in 2003, he worked from 2001 to 2008 in the studio of the architect António Madureira, participating in projects developed in partnership between architect António Madureira and architect Álvaro Siza. At the same time he developed his own projects. In 2005 collaborated with the publishing house DAFNE as a coeditor. Since 2006 he has been responsible for the course of Geometry at the EAUM. In 2002/2003 he worked as Monitor at the course of Design II at FAUP.

Eliseu Gonçalves

Faculdade de Arquitectura da
Universidade do Porto (FAUP)
Via Panorâmica S/N
4150-755 Porto PORTUGAL
egoncalves@arq.up.pt

Keywords: Orson Fowler,
houses, design analysis,
geometry, octagon, abstract and
organic, domestic architecture

Research

The Octagon in the Houses of Orson Fowler

Presented at Nexus 2010: Relationships Between Architecture and Mathematics, Porto, 13-15 June 2010.

Abstract. Orson Squire Fowler (1809-1887) was an American author who wrote a fantastic book about octagonal houses in the mid-nineteenth century. This present essay focuses on how radial geometry can be used as a tool placed for designing comfortable, affordable housing. Octagonal geometry can be used as a tool for controlling nature, or as a system for controlling construction. This duality is synthesized in Fowler's use of the octagon. The analysis is extended to two other buildings with octagonal plans, one ancient, the Hellenic "Tower of the Winds," the other contemporary, Álvaro Siza's "Mickey Mouse House."

Introduction

It is the loveliest study you ever saw ... octagonal with a peaked roof, each face filled with a spacious window ... perched in complete isolation on the top of an elevation that commands leagues of valley and city and retreating ranges of distant blue hills. It is a cozy nest and just room in it for a sofa, table, and three or four chairs, and when the storms sweep down the remote valley and the lighting flashes behind the hills beyond and the rain beats upon the roof over my head – imagine the luxury of it.

Mark Twain[1]

This present study focuses on the architectural thinking that Orson Fowler formulated in the book *A Home for All or The Gravel Wall and Octagon Mode of Building New, Cheap, Convenient, Superior and Adapted to Rich and Poor* [1848, 1853]. This treatise presented a persuasive array of practical actions that established a "community" of octagonal buildings still visible today, mostly in the eastern United States.

Outlining, although briefly, a kind of shape "genealogy" generated by the octagonal diagram allows us to locate the philanthropist and dilettante Fowler's proposal as an intermediate between such works as the Tower of Winds, built in Athens around the year 50 B.C. by the Greek Andronicus of Cyrrhestes, and the modern-day Pego Guesthouse in Sintra, Portugal, nicknamed the "Mickey Mouse House" designed by Álvaro Siza with António Madureira. The time span presupposes a systematic recourse to a particular geometrical form as the generating source of an order determinant for its rationality.

The example of these two singular buildings enables us to evoke some paradigmatic circumstances in the use of geometry as a system of formal, functional, technical and design control. In particular, on one hand, octagonal geometry as a tool for controlling nature; on the other, octagonal geometry as system for controlling construction. This duality is crucial for understanding the significance of Fowler's ideological and architectural proposal.

Nexus Network Journal 13 (2011) 337–349   NEXUS NETWORK JOURNAL – VOL. 13, No. 2, 2011   337
DOI 10.1007/s00004-011-0070-8; *published online* 8 June 2011

The Tower of Winds: the urgency to rationalize nature

The Tower of the Winds is mentioned in Book I of Vitruvius's *Ten Books of Architecture* and was the object of considerable attention in the Beaux-Arts academies of the late eighteenth century, thanks to the success of the drawings included in the book *The Antiquities of Athens* [Stuart and Revett 1787] (figs. 1 and 2). The Greek octagonal tower was built for reading the sun and the winds, in addition to hosting a monumental water clock powered by a water tank attached to one façade. This clock guaranteed the reckoning of time regardless of whether or not weather conditions made it possible to read the sundial. The construction was praised as a precision instrument for identifying the various winds.

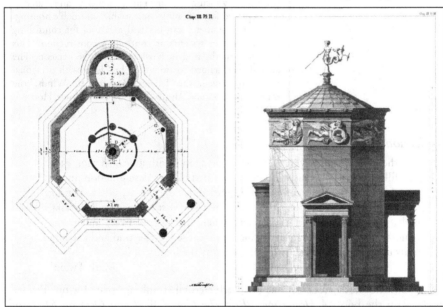

Fig. 1 (left). Tower of the Winds, plan [Stuart and Revett 1787: Ch. III, Pl. II]
Fig. 2 (right). Tower of the Winds, façade [Stuart and Revett 1787: Ch. III, Pl. III]

Vitruvius used the tower to warn about the problem of unhealthy cities. He said that the urban planning should take into account topography and its relation to the prevailing winds, which "if cold, are unpleasant; if hot, are hurtful; if damp, destructive" [Vitruvius 1826: I, VI, 24]. Thus, to determine the winds' orientation, it was necessary was to place a marble slab with a gnomon in the centre of the future city, on which could be traced the shadow of the sun in order to establish the cardinal directions; then a circle divided into eight equal parts would establish the direction of the eight winds. After describing the geometrical rules, Vitruvius suggests:

> Divide the remainders of the circumference on each side into three equal parts, and the divisions or regions of the eight winds will be then obtained: then let the directions of the streets and lanes be determined by the tendency of the lines which separate the different regions of the winds. ... Streets or public ways ought therefore to be so set out, that when the winds blow hard their violence may be broken against the angles of the different divisions of the city, and thus dissipated [Vitruvius 1826: I, VI, 27].

The Hellenic construction is in fact a cosmic machine and essentially represents a kind of scientific control over nature: its geometry tends to organize natural phenomena systemically. Vitruvius's transposition of geometry to city planning expands on this fact, using it to govern the construction of a new organization: in this case, urban morphology. This is an action of domination which, through a process of geometrical measurement and organization, leads to an abstract order. The structure of geometry thus provides a rationality that gives rise to an abstract built environment in opposition to the chaos of nature. As Ortega y Gasset said, the Greeks were the first ones to extend a "scientific reticular" over the natural phenomena and, therefore, put order in the "original confusion" [Ortega y Gasset 1995:123].

The Mickey Mouse House: abstract or organic

Regarding the advantage of geometry when it is subject to recognizable and rational rules, Wilhelm Worringer (1881-1965) mentions the elimination of the last life residue:

> We therefore put forward the preposition: The simple line and its development in purely geometrical regularity was bound to offer the greatest possibility of happiness to the man disquieted by the obscurity and entanglement of phenomena. For here the last trace of connection with, and dependence on, life has been effaced, here the highest absolute form, the purest abstraction has been achieved; here is law, here is necessity, while everywhere else the caprice of the organic prevails [Worringer 1997: 20].

This thought, first expressed in 1908, opened a discussion about the organic and abstract meaning, a debate which was of great significance to Modern Movement. On one side there was the abstraction of the geometrical forms with recognizable structure, product of a rational culture; on the other side, the organic, with biological links, which maintains the emotional and ontological connection between man and nature, recalling its position relative to the mechanics of the natural world.

The dialectic between abstract and organic forms can be observed in several of Álvaro Siza's works. For example, in the Pego House,[2] the octagonal "house-object" is contrasted with the contiguous "house-landscape" (figs. 3-5).

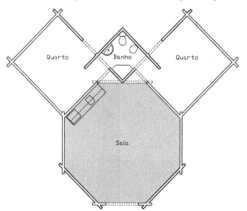

Fig. 3. Pego Guesthouse, nicknamed "Mickey Mouse," ground floor plan
(António Madureira Personal Archive)

The organically designed structure of the main house – irregular and apparently arbitrary – differs from the monolithic accuracy of the formal rigidity of the small volume of the guesthouse.

It is possible to identify two attitudes in the project linked to different architectonic languages[3] placed in close proximity: the spontaneous, sinuous and apparently unordered form, where "organicism" is augmented by the use of a wooden surface coating the façade; and the regularity and repetition of architectonic elements that emphasize the solitude of the octagonal house.

Fig. 4. Mickey Mouse House, Sintra, 2007 (Author's Personal Archive)

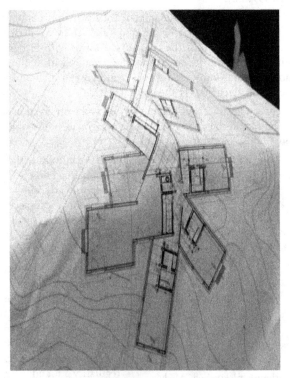

Fig. 5. Pego Main House, ground floor plan (Author's Personal Archive)

To use Christian Norbert-Schulz's (1924-2000) terms, the high degree of "concentration" [Norberg-Schulz 1966: 136] ascribed to the octagonal prism means that it is only able to accept expansion by aggregating other, autonomous elements. This condition leads to the independence of the pieces attached and a topological isolation. The desire and use of simple geometrical forms makes it possible for us to to put some concepts already present at the very beginning of the Modern Movement in direct relation to each other, particularly those related to categorization and standardizing. Although it doesn't make sense in the Sintra guesthouse to talk about a mass produced solution, the use of standard solutions was important for the building design process to ensure the rapid speed of execution and low cost of construction. The problem of submitting to criteria such as industrial design, pre-fabrication, typical elements or prototypes was solved by the capacity of the regular octagon plan to synthesize these requirements.

These two distinct uses of the octagon figure allows us to focus on the limits of Orson Fowler's proposal – the relation with the natural order and the creation of an abstract legitimacy – as well as its ability to solve different kind of problems in a single logical system.

The octagonal geometry of Orson Fowler's houses

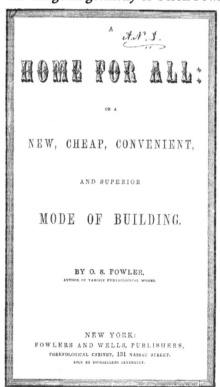

In 1848, Orson Fowler published the book *A Home for All: or a New, Cheap, Convenient, and Superior Mode of Building* (fig. 6).[4] The work is a manual of practical construction; the numerous re-editions it enjoyed during the second half of the nineteenth century attest to its influence and popularity.

Fowler was the most important advocate and practitioner of phrenology: now a pseudoscience, it was the precursor of modern psychology and neurology.[5] An idealist and man of action, his conception, focusing on "phrenological reason," eloquently addresses some of the social problems of industrial society and, particularly, Victorian age prejudices related to the emancipation of women, sexual education, the conjugal relationship in marriage, family organization and, therefore, the transformation of domestic space.

Fig. 6. Cover of Orson Fowler's *A Home for All: or a New, Cheap, Convenient, and Superior Mode of Building*, 1848

In Fowler's view, man's ability to build is a latent quality of thought because it is a primary need. Particularly, the revelation of man's "Inhabitiveness" and its "Constructiveness" [6] – to know how to Inhabit and to know how to build – gives him a special flair for designing his own home. This capacity allows him to develop his thinking about an "ideal house," which must be based on the octagonal panoptical geometry.[7]

Besides its historical significance as sacred geometry to structure religious architecture, namely those dedicated to the Virgin Mary and the Saviour, the octagon also seems to have a remarkable aptitude to support domestic programs. Before Fowler, we can identify other cases: Wadstrom proposal for temporary houses during the colonization of Africa [Wadstrom 1764]; Inigo Jones's octagonal house stated by William Kent in 1727 [Kimball 1922]; Thomas Jefferson's sketches of Octagon Houses (ca. 1800) and his famous Poplar Forest, Virginia (1809) [Fletcher 2011].

However, the model proposed by Fowler is drawn from outside the tradition that elects geometry as a mean of approaching the symbolic or the control of certain spatial effects. The universe in which Fowler moves is more prosaic and pragmatic. His scholarship is primarily an eclectic self-education that allows him, through the logical-deductive thinking stimulated by Phrenology, to synthesize and systematize some problems related to the conception, production and use of buildings in the social and economical context of the time.

Fowler's ideal form

The fourth chapter of the book *A Home for All*, entitled "Superiority of the Octagon Form," opens eloquently with the following question:

> But is the square form the best of all? Is the right-angle the best angle? Can not some radical improvement be made, both in the outside form and the internal arrangement of our houses? Nature's forms are mostly SPHERICAL [Fowler 1853: 82].

The problem of the ideal form is not found in formulas and geometric rules anchored in the classical academic tradition. The proposition is as follows:

> ... [because] the octagon form is more beautiful as well as capacious, and more consonant with the predominant or governing form of Nature – the spherical – it deserves consideration [Fowler 1853: 88].

Why does Fowler use the octagonal shape to support his thinking about a new "architecture of the house," contrary to the very tradition he himself calls the "Doric Style"? As we will see, the title chosen for the 1853 revised third edition sums up the reason for his preference: *A Home for All or The Gravel Wall and Octagon Mode of Building New, Cheap, Convenient, Superior and Adapted to Rich and Poor.*

To the question asked in the preface of the book – "Why so little progress in architecture, when there is so much in all other matters?" – Fowler replies with formal design austerity, with a new functional rationality and with a program of technical innovations, unusual for the times.

The house he built for his family in Fishkill, New York, illustrates and justifies the formal, technical and ideological proposals that the book outlines (fig. 7).

In keeping with the ideology of the *new man*,[8] the octagon is the accomplice of an architecture rooted in the practical sense of life and the promotion of a hygienic and comfortable environment. Surprisingly, in the nineteenth century this will make it possible to support a functionalist view of architecture based on certain basic principles: health criteria in the implementation, simple rules for circulation, space and technical functionality, establishment of environmental control mechanisms, updating and rationalization of materials and construction processes.

The octagonal diagram accommodates and orders this complexity, justifying it as an ideal model placed in the service of its doctrinal expansion.

occupied little during sunshine, but mostly evenings, and on special occasions; whereas living-room is used early and late, summer and winter.   One often needs to lotch in sunshine, and sitting-room is its place.   This a bay window, facilitated by the octagon form, promotes.

A COOL SOUTHERN BREEZE always accompanies right hot weather. This renders your sitting-room the coolest in the house, except those

20-FEET AND CENTRAL STAIRWAY PLAN.

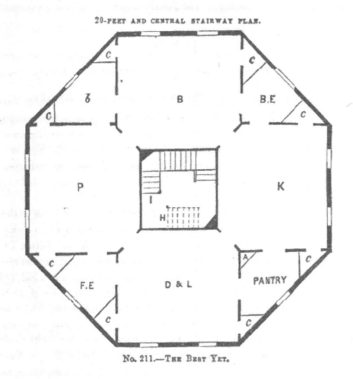

No. 211.—THE BEST YET.

right above it, whilst in fall, winter, and spring you want all the sun you can get in your *sitting*-room, even though it robs the others.

AN EAST OR WEST entrance will enable you to put your parlor on the north and sitting-room on the south side, while a northern entrance naturally gives the sun to the kitchen, and a southern to your parlor. These facts are worth considering in laying out the house you are to live in always, yet have heretofore remained unnoticed.

WOMEN SHOULD TURN ARCHITECTS.   They are naturally adapted

Fig. 7. Ground floor plan, Octagon House built in Fishkill, ca. 1848 [Fowler 1873: 1179]

Fowler starts his argumentation with a harsh criticism of the spatial organization and ostentatious ornamentation of the traditional American house. The metamorphosis of space suggested by Fowler is based on the principle of economy of means to provide a comfortable home accessible to everyone.

The persuasive strategy to establish his ideal is directly addressed to the matter of why the octagon should be chosen. Through a discourse based up on simple mathematical language and linear arithmetic relations, Fowler begins by comparing the efficacy of "compactness" determined by the ratio of areas between covered spaces and façades required for space enclosure (fig. 8). By implication, this will also correspond to the circulation areas required for the rooms. Because of its efficient correspondence between perimeter and area, the circle was deemed the perfect shape (fig. 9).

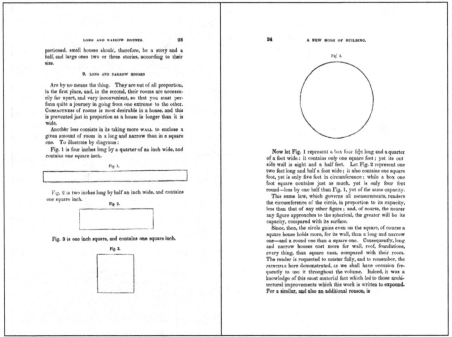

Fig. 8 (left). Geometric justification for the octagon plan: compactness [Fowler 1853: 23]
Fig. 9 (right). Geometric justification for the octagon plan: perfection of the circle
[Fowler 1853: 24]

The primacy of the circular plan thus promised to provide an alternative to the great bourgeois houses, with equal or greater comfort and at less expense.

With the cylinder as a conceptual framework, Fowler continues listing the reasons why the two most common housing types, the "winged house" and the "cottage," should not be built. Regarding the irrationality of the winged house he states:

> Wings on houses are not in quite as good taste as on birds. How would a little apple or peach look stuck on each side of a large one? Yet winged houses are just as disjointed and out of taste. … [L]et purse-proud, empty-headed nabobs throw away themselves, their comfort, and their Money on winged houses, but give me some other form [Fowler 1853: 74].

The winged house was problematic in terms of the comfort of space: the grandeur that was outwardly manifested corresponded to small and tight spaces inside; there were serious functional problems (for example, the "parlor" was too far from the kitchen); the radiant heat dissipated without creating a warm environment; the extension required to so that every room could be included in a single floor meant an increase in direct contact with the soil, increasing humidity problems, which was especially critical in the bedroom areas; the large surface area increased all thermal problems. All of these problems and others were used to justify the compact form as a matrix for establishing a new detached house solution.

Fowler also criticised the traditional English country house, which he calls "Doric style." The use of this typology essentially serves as a means to introduce the problem of ornament into his discourse.

> And here let me develop the law which governs this whole subject of taste and beauty. Nature furnishes our only patterns of true ornament. All she makes is beautiful, but, mark, she never puts any thing exclusively for ornament as such. She appends only what is useful, and even absolutely NECESSARY [Fowler 1853: 75].

One of the architectural issues that would be discussed at the turn of the century was related to a domesticity that is either in tune with the natural world, or in opposition to it, assuming an abstract form to be more consistent with the phenomenon of the mechanization of society. Fowler anticipated the theoretical bickering that would be conducted in the beginning of the twentieth century by critics like Adolf Loos (1870-1933). Regarding the utility of the absurd "cottage" with its multifaceted form and inclined roofs, he said:

> The BEAUTY of a house is scarcely less important than its room. True, a homely but CONVENIENT house is better than a beautiful but incommodious one, yet beauty and utility, so far from being incompatible with each other, are as closely united in art as in nature. ... beauty and utility are as closely united in architecture as they are through out all Nature. ... Form embodies an important element of beauty [Fowler 1853: 87].

Concerning the differences between the circle and the square shapes, Fowler discusses the problem of dead spots in the square plan (fig. 10). The spaces formed by planes arranged perpendicularly suffer from the non-use of these corners. The defects are categorical: the amount of material and labour expended unnecessarily to build them; it is more difficult to heat, ventilate and light these spaces; they impose constraints on the provision of furniture. It was here that the sphere hit upon the perfect order:

> Nature's forms are mostly SPHERICAL. She makes ten thousand curvilinear to one square figure. Then why not apply her forms to houses? Fruits, eggs, tubers, nuts, grains, seeds, trees, etc., are made spherical, in order to enclose the most material in the least compass.

> ...Why not employ some other mathematical figures as well as the square? These reasoning developed the architectural principle claimed as a real improvement, and to expound which this work was written [Fowler 1853: 82].

In a comparative analysis between natural objects and the sense of beauty of form, Fowler establishes a clear relationship between Beauty and round shapes as a universal principle. Therefore on a scale of values the triangle is at the lowest level and the circle at the highest. For Fowler this is an immutable law of nature, and he contrasts the beautiful shape of a dome with the sharp lines of the "cottage." Thus, by approaching the perfection of natural order, represented by the circle, the intermediate octagonal shape provides the geometric matrix that encapsulates the benefits of the Octagon House (fig. 11).

Given the superiority of the octagonal plan, only a historical doubt remains: Why haven't octagonal houses been successful in the past? Fowler attributes this to the use of wood. Wood was the material of choice during the expanionist period in the United States for two complementary reasons: it was readily available and abundant; it was easier to work with than stone masonry and didn't require skilled labour. Thus, except for institutional buildings, the foundation and urban development of cities was done using lightweight constructive solutions in wood that evolved into the typical "balloon frame." Now, this way of building favoured right angles; this is the reason that Fowler cites for the lack of interest in the circular plan in general and octagonal plan in particular.

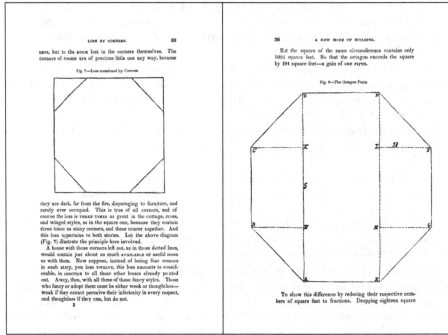

Fig. 10 (left). Geometric justification for the octagon plan: octagon vs square
[Fowler 1853: 33]
Fig. 11 (right). The octagon as a mediating geometrical figure [Fowler 1853: 36]

The solution he proposed was the "gravel-wall" system. As stated in the title of the revised third edition of his book, the buildings must be constructed with a "gravel-wall." The affirmation is evidenced in several passages of the text; the construction was to be based on peripheral walls built using a kind of concrete that could use local rocks. He traces the discovery of this solution back to an episode in 1850 in Jaynesville (Wisconsin), when he met Joseph Goodrich [Fowler 1873: 20]. The discretion of the system coincides with the formulation presented in 1836 in London by George Godwin (1813-88). Godwin is regarded as the inventor of modern concrete (without steel); his "Essay on the nature and properties of Concrete, and its application to construction, up the present period" (1836) earned him that year's prize of the Institute of British

Architects of London. To Fowler, the discover of such material was crucial to the recovery of the octagonal building and, therefore, to the construction of a new "modern" and "democratic" housing solution. For this propose, the radial geometry was beneficial because permits technical innovations and creates a rational distribution of space, structure and infrastructures;[9] meanwhile, its panoptical quality allows a new kind of control over domestic movements.

Although the discussion of the advantages of the spherical shape finds a parallel in the utilitarian thinking of the classical academies of the late eighteenth century,[10] the sense that the American phrenologist gives it constitutes a unique moment in the context of domestic architecture in the Victorian age.

The success of the Octagon House due to a hybrid description that ranges from the scientific to the empirical, makes explicit an analytical and operative rationality that could be illustrated by examining the daily reality of the nineteenth century. A simple statement of the problem and the appointment of a solution that was standardized yet adaptable, comfortable yet inexpensive, innovative yet built using available means, not only led to an unprecedented number of houses being built, it also triggered a process which recognizes in Fowler's proposal a "support," that is, to use the definition of the Dutch architect John Habraken (b. 1928), that which is structural, unchangeable and collective [Habraken 1972: 92-93].

Conclusion

In the wake of encyclopedists and naturalists of the eighteenth century, the practice of phrenology involved a process of cataloguing, selection and standardization of human behaviour intended to establish *types*. Fowler would transpose that same desire into the field of architecture in order to structure a simple and logical form whose ideal could be disseminated. The selection of the octagon as a functional and formal unit of a morphological type appears to emerge from a process analogous to that of "phrenological" rationality.

The model proposed by Fowler became a house archetype that would last for a century. The exotic nature of the octagonal became secondary in the pursuit of an architecture rooted in the practical sense of life, in comfort and in hygiene, in the relationship to territory and climate. Fowler transforms the octagon into an *apparatus* of a social ideal based on technical generosity disciplined by a regular geometry. This fact becomes particularly important for architecture, if we think of George Fred Keck's House of Tomorrow[11] and Buckminster Fuller's Dymaxion House, proposals built at the start of the Modern Movement and directly inspired by the octagon house.

The survival of the model is based on a geometrical justification that explains two guiding principles. The first is concerned with natural phenomena understood in a physical dimension (climate, topography, etc.), and also perceived as manifestations of a universal organizing law that rationalizes Beauty. The second matches a particular geometry to a technical skilfulness that makes it possible to create a type, and therefore the ability to adapt without losing the logic of form.

Notes

1. Mark Twain, in a letter to William Dean Howells, 1874.
2. The Pego House constructed in Sintra, by architects Álvaro Siza Vieira and António Madureira, was finished in 2007. It consists of the main house and an adjacent, smaller one, which served as a seasonal residence during the construction phase of the main residence; it

now is used as a guesthouse. The small pre-fabricated building was designed to be rapidly built and functionally pragmatic.

3. We can compare this with what had happened in The Hague (Netherlands) twenty years earlier. We refer to the twin houses "Punkt und Comma," built between 1983 and 1988; see for example, Alvaro Siza Vieira Progetti per L'Aja and J.D.Besh, "Elogio della transformazione", in *Casabella* 538, Electa, September 1987, pp. 4-15.They constitute a formal counterpoint between the functionalist rationalism and the expressionism; two aspects that had marked the history of modern Dutch architecture.

4. *A Home for All or The Gravel Wall and Octagon Mode of Building New, Cheap, Convenient, Superior and Adapted to Rich and Poor* [Fowler 1853] is the updated third edition of the book published in 1848. The 1853 edition will be the most widespread in the successive editions.

5. During the nineteenth century Phrenology – knowledge and classification of human behavior through "topographical" readings of the skull – gained social recognition as a form of psychological insight and personal growth, which would only come to vanish with the advent of psychology and neurology in the twentieth century. In Fowler's own words: "PHRENOLOGY, derived from two Greek words, [mind, or discourse, and treatise] consists in certain cause and effect relations existing between particular developments and forms of the brain, and their corresponding manifestations of the mind; thereby disclosing the natural talents and proclivities of persons from the forms, sizes, and other organic conditions of their heads" [Fowler 1873: 115] The distinguished patients Fowler tells of are Samuel Langhorne Clemens (1835-1910) – Mark Twain – and the writer Walt Whitman (1819-92). For example, in his work Whitman include some of the topics covered in phrenology; in turn, Fowler published some of Whitman's texts in his publications, namely, in the *American Phrenological Journal.*

6. "Inhabitiveness" and "Constructiveness" are part of the faculties of the mind relatable to certain areas of the brain: "Inhabitiveness" corresponds to "Species I - Domestic Propensities," while "Constructiveness" are part of the "Species II - semi-intellectual Sentiments" which are part of "Genius II - Human, Moral, and Religious Sentiments," both included in the Order I (Affective Faculties and Feelings) [Fowler 1840: 45-50].

7. The term used is derived from the work of English philosopher and jurist Jeremy Bentham (1748-1832), a Reformist. In *The Panopticon Writings* [1995], Bentham proposes a circular prison with a convergent organization of space in the centre. "Centre power," as a spatial value, was implemented by Orson Fowler in the Octagon House; the centre is the area of distribution, the lighting, essential both in terms of structure and management of infrastructure.

8. With specific reference to the hygienist idea of "healthy body, healthy mind" and a new status of the family based on a greater independence of the household.

9. For example, the Fishkill Octagon House had running hot and cold water supplied by gravity, central heating, roof top coverage with tanks to collect rainwater and subsequent filtering for consumption, artificial lighting with natural gas, and ventilation system by convection.

10. One of the paradigmatic examples in the environment of the Academy is the French architect Jean-Nicolas-Louis Durand (1760-1834). The tendency to produce efficient buildings by the affirmation of *utilitas* and *firmitas* will be transported by Durand to a rationality that separates beauty of form from its ornamentation. As stated by Peter Collins, the theory of Durand was particularly marked both by their connection to Ecole Polytechnique and by the financial situation of France which determined preferably utilities programmes [Collins 1977: 19].

11. The octagon houses of Orson Fowler were built over a widespread area; one of these houses was built in 1853 in Watertown, in a property that bordered upon the family of George Fred Keck (1895-1980). During his academic education in architecture, the octagon house would gradually go from being an object of curiosity to the subject of intense study. This will be one of the reasons why, in 1933, the radical functionalism and technical patent in the House of Tomorrow is repeated in a container similar to that Fowler had used fifty years.

References

BENTHAM, Jeremy. 1995. *The Panopticon Writings* (1789), Miran Bozovic, ed. London: Verso Books.

COLLINS, Peter. 1977. *Los ideales de la arquitectura moderna; su evolución (1750-1950).* Barcelona: Gustavo Gili

FLETCHER, Rachel. 2011. Thomas Jefferson's Poplar Forest. *Nexus Network Journal* **13**, 2 (Summer 2011).

FOWLER, Orson S. 1840. *Fowler's Pratical Phrenology.* New York: Fowlers and Wells.

———. 1848. *A Home for All: or a New, Cheap, Convenient, and Superior Mode of Building.* New York: Fowlers and Wells.

———. 1853. *A Home For All, or The Gravel Wall and Octagon Mode of Building.* New York: Fowlers and Wells (revised 3rd ed.) (rpt. New York: Dover, 1973).

———. 1873. *Human Science or Phrenology.* Chicago: The National Publishing Co.

HABRAKEN, John. 1972. *Supports: An Alternative to Mass Housing.* B. Valkenburg, trans. London: The Architectural Press.

KIMBALL, Fiske. 1922. *Domestic Architecture of the American Colonies and the Early Republic.* NEW YORK: SCRIBNERS (RPT. NEW YORK: DOVER, 2001).

NORBERT-SCHULZ, Christian. 1966. *Intentions in Architecture.* Oslo: Scandinavian University Books.

ORTEGA Y GASSET, José. 1995. *El Sentimento Estético da Vida.* Madrid: Tecnos.

REYNOLDS, Mark. 2008. The Octagon in Leonardo's Drawings. *Nexus Network Journal* **10**, 1: 51-76

STUART, James and Nicholas REVETT. 2008. *The antiquities of Athens* (1787). New York: Princeton Architectural Press.

TAVERNOR, Robert. 1991. *Palladio and Palladianism.* London: Thames and Hudson.

VITRUVIUS. 1826. *The Architecture of Marcus Vitruvius Pollio in Ten Books,* Joseph Gwilt, trans. London: Priestley and Weale.

WADSTROM, C.B. 1764. *An essay on Colonization.* LONDON.

WORRINGER, Wilhelm. 1997. *Abstraction and Empathy.* Michael Bullock, trans. Chicago: Ivan R. Dee, Inc.

About the author

Eliseu Gonçalves is a Portuguese architect and professor at the Faculty of Architecture of the Oporto University (FAUP), where he graduated in 1994. He has experience in urban planning and housing design; his work includes the Oporto Historical Water Front Renewal (with architect Manuel Fernandes de Sá) and a cluster of houses in northern Portugal. Since 1999 he has been teaching Building Construction at FAUP. The early academic work was concerned with the integration problem of engineering subjects in architectural courses: "The Language of Gravity – Observations on the Way of Teaching the Reasons Why Structures Work." His current interest is the issue of housing space transformation in relation to the new comfort hardware that became visible through the twentieth century in Portugal. He has finished the Architectural Ph.D. Course (PDA/FAUP) and is writing a Ph.D. thesis related to popular housing in Oporto between the periods of Monarchy and the Dictatorship. He is also member of the research group "Atlas da Casa" (House Atlas) in the Architecture and Urban Design Investigation Centre – CEAU/FAUP.

Teresa Marat-Mendes[*]

*Corresponding author

Lisbon University Institute – ISCTE
Av. Forças Armadas
1649-026 Lisbon PORTUGAL
teresa.marat-mendes@iscte.pt

Mafalda Teixeira de Sampayo

Lisbon University Institute – ISCTE
Av. Forças Armadas
1649-026 Lisbon PORTUGAL
mafalda.sampaio@iscte.pt

David M.S. Rodrigues

Lisbon University Institute – ISCTE
Av. Forças Armadas
1649-026 Lisbon PORTUGAL
david@sixhat.net

Research

Measuring Lisbon Patterns: Baixa from 1650 to 2010

Presented at Nexus 2010: Relationships Between Architecture and Mathematics, Porto, 13-15 June 2010.

Abstract. The present study, based on a comparative analysis of several plans for Lisbon's Baixa district, with an emphasis on that area's public space, contributes to an understanding of the urban design process and presents a fresh perspective on dealing with historical data by conducting a posteriori analysis using mathematical tools to uncover relations in the historical data. The nine plans used were quantified and evaluated in a comparative manner. While CAD was used to quantify the urban morphology of the different plans, comparative tables make it possible to register the data, which was further evaluated through two interrelated processes: mathematical analysis and the urban analysis. The results show the existence of power law relations for the areas of each of the city's different elements (e.g., blocks, churches, *largos* and *adros*). We discuss how this contributes to the understanding of the plans' elements.

Keywords: urban design, design analysis, design theory, CAD, computer technology, morphology, patterns, proportion, measurement, mappings, geometry analysis, fractals, graph theory, statistical analysis

1 Introduction

A city study cannot be separated from the ways in which people live in the city. The ways that people make use of public space are not totally predictable at the architect's drawing board. However, although the reactions of a city's population to its urban complexes might be unpredictable, some almost universal rules exist that allow architects to draw the contemporary city and these rules are already present in the cities of the Enlightenment. Today, thanks to the evolution of science, it is possible to create virtual systems that allow us to project and explore the possible outcomes of the population's reaction to a specific public space. The reaction of populations is then a result of the interplay between different components of the city. From this interplay, the complex dynamics of the urban form emerges.

Having these concerns in mind, the present paper will evaluate the Baixa area of Lisbon, which has been subject to several plan proposals from the period of the Enlightenment time up to the present day. Furthermore, this paper will examine how this public space has performed over time.

2 State of the art

The nine plans that are analyzed in the present paper refer to four different historical periods: Renaissance, Enlightenment, Modern, and Present-day. Nevertheless, the plans of the Modern period and the Present-day are based upon the chosen eighteenth-century plan.

In the eighteenth century, Portuguese Prime Minister Pombal demanded that Manuel da Maia, the Portuguese Kingdom's principal engineer, to establish rules for the reconstruction of the city of Lisbon after the 1755 earthquake [Sampayo and Rodrigues 2009; Marat-Mendes 2002]. To accomplish this, Manuel da Maia prepared a three-part dissertation indicating five ways to reconstruct the city. The selected approach recommended that the Baixa reconstruction should be situated exactly in the same place, but according to a new plan.

Two sets of rules for the reconstruction of Baixa were established in the third part of Maia's dissertation: One rule obliged the plan proposals to localize the churches exactly as they were before the 1755 earthquake and a second rule granted freedom to the plans' proposals regarding the location of the churches [Sampayo and Rodrigues 2009].

The plans for Baixa have been exhaustively analysed by researchers in different fields, including historians, architects, urbanists, and lawyers. Nevertheless, their research studies have been performed in an isolated manner. For example, França [1987] and Tostões and Rossa [2008] have analysed Baixa from an historical perspective. Heitor [1999] and Kruger [1998] applied a space syntax analysis to evaluate the urban form of Baixa. Mullin [1992] used an historical and political analysis to show the social mark that would be made by the newly restored city of Lisbon. Moreira [1993] aimed to define a conservation proposal for the Baixa and develop a replicable method of analysis for urban conservation. Monteiro [2010] studied Baixa in terms of the legislation produced during the eighteenth century and its influence on subsequent urban legislation. Lopes dos Santos analysed the construction technologies used to renovate Baixa and its relevance is shown [1994] by his analysis of the urban form.

Literature on the analysis of urban form also embraces diverse fields. The following section is divided into two main parts. The first part discusses the analysis of urban form from historical, physical, and morphological perspectives, while the second part addresses the mathematical analysis of the elements of urban space.

The analysis of urban from

a) The historical and physical perspective

Research into the development of cities during different historical periods constitutes an important contribution to the analysis of urban form. Thus, it is important to stress that the use of historical analysis does not represent an intention to replicate the past, but rather to provide the foundation for both preservation and innovation. Understanding the urban arrangements that have worked well, or that have produced beneficial effects in the past, provides important lessons for the future.

According to Marat-Mendes [2002], people's interactions with their environment not only constitute an area of research interest for sociologists or historians, but is also a growing area of interest for researchers of urban form. This is evidenced in Rapoport's [1990] explanation of people's interactions with their environment, Kevin Lynch's [1981, 1996] studies of people's images of cities, and Moudon's [1986] analysis of people's interactions within the built environment in residential San Francisco. Furthermore, the analysis of environment behaviour has also been enriched by Rossi's [1983] analysis of the processes shaping the urban environment and the work of brothers Léon and Rob Krier [L. Krier 1979; R. Krier 1979], who favoured the re-creation of an eighteenth and nineteenth century grid-planned European city by focusing their rhetoric on the importance of tradition. Indeed, this search for an historical understanding of how urban fabric is shaped over time, and in a comparative manner, is reinforced in the present work's argument for the need to obtain a better understanding of the common processes that have shaped this urban fabric.

However, from the works identified as relevant for the study of urban form, it is evident that different authors use different methods. Papageorgiou's book, *Continuity and Change. Preservation in City Planning* [1971], reveals an interesting analysis of historical centres, while stressing the importance of rehabilitation and conservation. It also recognizes the value of cultural heritage for the survival of historical urban centres. Of particular relevance to the present work is Papageorgiou's acknowledgement of the importance of reading urban space through its multi-layered urban formations. Moreover, he identified the occurrence of change in the urban fabric as the way in which historical urban centres become lively urban spaces with a special kind of atmosphere, i.e., places that provide quality of life to their inhabitants.

Moudon's *Built for Change* [1986] presents a specific case study that analysed the evolution of a residential neighbourhood around Alamo Square in San Francisco. This study addresses the history of that neighbourhood, from its origins in the nineteenth century up to the 1970s, by examining the interactions between different elements of urban form. While making use of historical sources to identify the historical patterns that shaped that particular neighbourhood, the study also promoted an examination of the interaction between people and environment. According to a later study by Moudon, urban form can "only be understood historically since the elements of which it is comprised undergo continuous transformation and replacement" [1997: 7]. Although Moudon's and Papageorgiou's studies differ in scale and nature, they seem to find common ground in the methods they used to understand the different processes that shape urban form. These methods are based on an historical analysis that accepts that changes in urban form occur in multi-layered formations that continually undergo transformations and replacements.

Regarding the historical analysis of urban form and the processes of transformation and replacement that occur within it, Morris [1994], in his historical account of the history of urban form from its origins to the Industrial Revolution, describes the physical results of some 5000 years of urban activity. He constantly cites the socio-economic constraints or political issues that relate to the formation, alteration, or adaptation of the urban forms. Although Morris's analysis is very extensive in terms of the number of cases analysed, he does not provide a comprehensive analysis for each case. Nevertheless, he provides important contributions to the perceptions of what, in this paper, is described as 'lessons from the past.'

Yet, the lessons from the past must not be misunderstood by a single reading of historical urban places. They should also include a re-examination of recent and implemented urban proposals that have proven to be successful. Thus, a historical survey should always be included in any physical analysis of urban form, either to learn about past successes or to understand the forces that have shaped the urban form over time.

The present work sees the dual process of analysis as essential to the determination of liveable and enjoyable urban places for its citizens.

b) The urban morphology perspective

In her comparative analysis of three historical examples of planned urban development, Marat-Mendes [2002] aimed to identify some practical reasons for sustainability. The historical examples cited included the urban development of Lisbon's Baixa by Eugénio dos Santos and Carlos Mardel in 1756, the development of Edinburgh's first New Town by James Craig in 1767, and the development of Barcelona's Ensanche by Ildefonso Cerdà in 1855. Marat-Mendes's work has contributed to the analysis of urban form and its transformations by identifying interest in the perspective of urban morphology, namely by acknowledging the work of geographers such as Conzen [1960] and Whitehand [1981, 1990, 2000, 2001] who contributed to the refinement of morphological studies, with specific analysis on British medieval cities.

According to Whitehand [1981: 17], Conzen's most significant contribution rests on the fact that the evolution of the urban fabric results from progressive physical processes. However, to identify these progressive physical processes, a need exists for a "detailed examination in genetic terms of street and building lines, building block plans, and the shape, size, orientation and grouping of plots (lots)" [Whitehand 1981: 17]. Carmona et al. [2010] state that Cozen has worked these components, while considering them as key elements in the analysis of urban form.

Conzen's and Whitehand's morphological studies have been utilised in other disciplines, such as in Samules' [1993, 1997, 1999] works on architecture. Urban morphology analysis has also been followed with interest in Italy, where it was first approached by Muratori [1959], giving rise to new areas of urban research, such as urban typology. Muratori was one of the principal initiators of urban typology and renamed it Procedural Typology. This consisted of a morphological analysis of towns where special attention was given to building types as the elemental root of the urban form and where evolution was the main topic. The "notion of evolution was therefore invoked as a means of gaining a better understanding of the built environment" [Kropf 1998: 45].

In the late 1960s in France, the interest in morphology and typology attracted architects Phillipe Panerai and Jean Castex and the sociologist Jean-Charles DePaule. However, their contribution to building typology was less important than their contribution to the deeper analysis of the city block [Castex et al. 1977].

Other contributions to the morphological analysis of urban forms have focused on the evolution of street layouts and block form. In this area, Sola Morales and the 'Grupo 2C' have contributed to urban block analysis, devoting special attention to the time between the medieval period and the nineteenth century. Their principal case study focused on Barcelona's urban fabric. In 1978, *Lotus International* dedicated an entire issue to this specific subject. Issue no. 19 of this journal included several studies of urban blocks, such as the Sola-Morales' [1978] study of the evolution of the blocks in Barcelona, and Léon Krier's article, "Fourth Lesson: Analysis and project of traditional urban block" [1978]. Krier's study underlined the importance of the return to traditional urban patterns of blocks, streets, and squares as fundamental elements of urban form. He proposed the analysis of these urban elements based on historical or 'traditional' models of the European city.

Not only has the return to traditional urban patterns and sizes been claimed as essential to the urbanisation of urban spaces, there are also indications that the way urban forms are analysed has changed and need to be reoriented towards traditional methods. According to Siksna,

> A further limitation of the recent studies and applications of the block is that they have confined their attention to the relationship of built form, street space and internal courtyard space. Most block developments are conceived, and carried out, as comprehensive schemes for the entire block, rather than the traditional manner of individual buildings on separate lots. Thus the block has been revived essentially as an urban form of buildings and spaces, but not as a layout form of lots, blocks and streets [Siksna 1990: 2-8].

Thus, Siksna gives preference to the analysis of urban forms by using the urban block as a key unit of analysis. He claims that the practice of conceiving urban blocks needs to be rethought and he proposes an analysis of the traditional methods of block arrangement.

So far, this review of studies on the physical analysis of urban forms has focused on traditional urban patterns and the preferred methods of urban analysis. Therefore, both perspectives seem to agree that the urban fabric needs to be analysed using historical methods because the city cannot be studied separately from the processes of change to which it is subjected [Moudon 1997].

Despite the existence of a significant number of works on analysis of urban form, the relationship between urban elements and the processes that enabled change appears to be an area of investigation that has not been fully explored. According to Moudon:

> … the rate of change in either the function or the form of the cells varies from city to city, but also generally fits into cycles related to the economy and culture. Building and transformation cycles are important processes to explore for city planning and real estate development proposes, yet are rarely studied in contemporary cities [1997: 7].

This lends support to the validly of the use of the comparative analysis of urban form in this work, wherein the same area of the city is studied in reference to different historical periods.

Mathematical analysis of urban spatial elements

The study of the city can be conducted at different levels by abstracting features and using different tools. Graph theory was first applied to solve an urban problem in 1735, when Euler [1741] used it to solve the problem of the Königsberg bridges. In the 1980s and 1990s, space syntax revitalized the use of the graph theory for measuring city features [Hiller and Hanson 1989]. One aspect of this theory, of particular interest to this study, is its use of the term "iosvist" to define the volumes of space seen from a point in the city [Benedikt 1979]. More recently, Agent-Based simulation has gained particular interest because it is not possible to account for some non-linear features with traditional reductionist approaches [Batty 2007]. The non-linearity of the social aspects of life systems is also manifested in cities and the mathematical analysis of urban spatial networks and can be seen in Blanchard and Volchenkov's [2008a, 2008b] work on random walks. Random walks defined on graphs, and their very closely related diffusion processes, have been studied in detail. The random walk hypothesis was used successfully in several fields of research such as economics, where it was used to model share prices, and population genetics, where it was used to model genetic drift [Blanchard and Volchenkov 2008b].

Miller [1978] compared the city with a biological entity that shares several physiological characteristics with biological systems scale, such as the mass of their bodies (*M*). The power P_w required to sustain a living organism has been shown to scale according a power law, such as:

$$P_w \propto M^{\frac{3}{4}} \qquad \qquad \text{(eq. 1)}$$

This characteristic (features obeying some power law distribution) has also been observed in many aspects of cities [Savage and West 2006]. Zipf [1949] showed that the population of a city depends only on the size of the largest city and the rank of the city.

In effect, power laws have been found in several domains of human activity. From the World Wide Web to genetic pathways, this law, which affects all scales of observation of the objects, is pervasive and constitutes a fundamental feature of human activities.

In this sense, we analysed the areas of different elements of Baixa, including the block area, the church area, and the *largos* e *adros* areas. We found that these areas also obey characteristic power laws.

3 Case study and methodology

Case study

In this paper, the case study analysis refers to Baixa, Lisbon's downtown area. Several plans were used in order to make a comparative analysis possible, including the pre-earthquake plan by Tinoco (1650), the plans 1 (Pedro Gualter da Foncêca and Francisco Pinheiro da Cunha), 2 (Elias Sebastião Poppe and Jozé Domingos Poppe), 3 (Eugénio do Santos de Carvalho e António Carlos Andreas), 4 (Pedro Gualter da Foncêca), 6 (Elias Sebastião Poppe), and the "Chosen" plan[1] (missing), proposed for the reconstruction of the downtown in the aftermath of the 1755 earthquake (1758). We also considered the E. Gröer 1948 proposal for the transformation of Baixa's urban blocks, and the built or Present-day plan for Baixa (2010).

Concepts

This research has considered Public Space as the sum of two main urban morphological elements. One use of the term public spaces refers to the circulating spaces (streets area) and the other use refers to the permanence spaces[2] (squares, churchyards or *Adros* and *Largos*) [Pereira 1983]. Thus, for its analysis, this research will consider the following principles:

- 'Public Space' – The sum of Streets, Squares, and *Adros and Largos* areas
- 'Permanence Space' – The sum of Squares, and Adros and Largos areas

A short definition of the Permanence Space terms is succinctly provided as follows:

- 'Squares' – A geometrical space, bordered with houses or rivers on each side, usually with public buildings on these edges.
- '*Largos*' – A Portuguese term used to designate an informal square.
- '*Adros*' – A Portuguese term used to designate an open space situated in front of a church.
- 'Residual Voids' – Urban Voids that cannot be classified as public spaces and that do not have any useful purpose.
- 'Block Area' – The sum of the built area and the inner courtyard area.

These definitions, although useful, are insufficient for grasping the concepts in their entire scope. The interplay between these elements is richer and more useful. The following explanations attempt to elucidate a deeper understanding of their relations.

Squares refer to public spaces with a well-defined (Euclidean) geometry, which result from the addition or subtraction of different volumes, and which present different uses that respect the use of a square.

The square is a public space *par excellence*, as are the street, the avenue, and the alley. The street acts as a flowing space while the square acts as a space of interaction. Thus, the square functions as a space of permanence associated with the functions that are present within it. The *largo* differs from the square. Therefore, the *largo* is generally the result of the exploitation of a residual urban space and presents mostly irregular forms with diverse sizes. The *largo* might emerge from the need for a void or a crossway. However, it does so in a spontaneous form and not in a previously planned form as would occur in a square. The square's space is rigid and defined by rules; it has functional order constraints while the purposes of the *largo* are not as clear [Sampayo 2007].

The consolidation of the square as the power headquarters of the city, as an ordered space and as a planned infrastructure, only emerged during the seventeenth and eighteenth centuries.[3] According to Sampayo [2007] the plans for medieval Portuguese cities obeyed the criteria used for regular tracing, but these plans did not include defined urban squares. Empty spaces existed near the city's walls but these were not structured. Today, if we find structured medieval squares within medieval cities, these are the result of prior transformations. During the reign of Portuguese King Manuel I, at the end of the fifteenth century and beginning of the sixteenth century, institutional reforms with urban concerns were instituted, particularly with regard to the creation of regular squares associated with the construction of new civil and religious buildings. However, the eighteenth century inherited this cultural past and it is during that time that the square, as a public space, was designed in accordance with a geometric structure and within this framework one can trace the evolution of public space within the city [Sampayo 2007].

Françoise Choay, an urban and architectural historian from the University of Paris – VIII claims that the history of the public square can be written throughout the history of urbanization and of power, revealing chronological discrepancies as well as morphological differences among different countries. Therefore, she divides her analysis [Choay and Merlin 1988] into three main stages. The first stage refers to the Medieval Era, from the eleventh century until the end of fourteenth century; the second stage refers to the period from the Renascence Era until the Industrial Era; and finally, the third stage refers to the era in which industrialisation flourished.

According to Choay and Merlin [1988], during the Medieval Era, very few squares existed in cities outside of Italy. In Paris, up to the reign of Henri IV only one square can be identified. Thus, public life used to take place in the streets. The tiny churchyards, located in front of the churches and cathedrals, have not, thus far, earned the distinction of being designated as squares.

Nevertheless, during the Medieval Era, it is important to identify two exceptions: the squares situated over the forums of the former Roman foundations; the arcaded squares that occupied several modules in the centre of the orthogonal grid and that usually joined a church and a commercial area in the city centre (as in France in Montpazier, Villeréal), and the squares in fortified cities and in new cities such as Zähringen in Switzerland. These examples testify to the existence of squares in planned medieval cities, known as Bastides, which were also established in Portugal.

During the second stage, and primarily due to Italian influences, an aesthetic square emerges. The aesthetic square did not serve a functional purpose. Instead, the principal propose of the aesthetic square was to embellish the city and convey an image of power. The aesthetic square was no longer a product of a municipality or a collective; instead, it was a final product of architects, promoters, and urban art.

The third stage is marked by the flourishing of the Industrial Era in which public life took place in the interior of public buildings; such as markets, theatres, etc. During this period, the public space was invaded by the means of transportation.

As previously noted, the concept of the square evolved over time and is strongly associated with specific periods of our history.

The notion of public space also emerged very late in time and has been dealt with in different ways by different authors of urban studies. Léon Krier [1999] envisioned the public space as a fraction of a specific urban environment that defines the lifestyle of the society that inhabits it. In this sense, Krier believes that public space should comprise between 25% and 35% of the overall plan of a well-determined urban design morphology.

The methodology

As referred to in the previous work of Sampayo and Rodrigues [2009], research in the Portuguese urban cartographic archives has revealed the existence of several copies of the same plans with subtle differences. The following four plans can be found in the "Gabinete de Estudos Arqueológicos da Engenharia Militar"(GEAEM), although França [1987] has indentified five plans during the 1960s, meaning that one of the plans is missing [Sampayo and Rodrigues 2009]: Plan 1, Plan 2, Plan 4, and Plan 6. The following plans can be found in the City Museum: Plan 1, Plan 2, Plan 3 (two identical versions), Plan 4, Plan 6, as well as the plan for downtown before the earthquake.[4]

Sampayo and Rodrigues [2009] already noted the existence of two or more copies of the same plans. They also indicate the existence of slight differences in these copies. However, ongoing research, led by Sampayo, is conducting a deeper analysis of this occurrence. Moreover, it is important to stress that the eighteenth-century cartography, analyzed in the present article, does, in fact, refer to the plans that are found in the City Museum in Lisbon.

The methodology adopted for this paper followed the following steps:

1. Delimitation of the area of study. Approximately the same area (51 hectares) was considered for all nine plans.

2. The plans were digitalized and rescaled to the same scale to allow for the comparison of components contained in the nine analyzed plans.

3. The digitalized plans were transformed into vector format in order to allow for the measurement analysis of their urban from elements. The analysis was realized using the AutoCAD tool.

   Seven of the nine analyzed plans refer to the eighteenth-century cartography that was digitalized from the City Museum original plans. The two other plans are the 1948 plan for Baixa by the French architect-urbanist Ètienne de Gröer, and the Present-day plan for Baixa. All nine plans were analyzed in a comparative manner through AutoCAD.

4. The comparative analysis was based on the following elements: churches, noble buildings, squares, urban blocks, and permanence areas.

5. Each of the different plans was then analyzed for the size distribution of each of the different components of interest: block areas, church areas, *largos* and *adros*. Furthermore, statistical data derived from the resulting features was compiled and further analyzed to provide insights into how these features interacted in the city urban form. From this, we obtained several explanative correlations.

4 Results and analysis

Area distribution of urban elements

We analyzed the size (area) distribution of the several urban elements included in the different Baixa plans, including block size areas, church areas, *largos* and *adros* areas. These three features were examined only for the chosen area of study. It was observed that some of the elements of the urban fabric obey a characteristic power law distribution in terms of occupied area. This means that the area of a feature only depends on the biggest feature of the same type and on the rank that it occupies according to:

$$A_n \propto A_1 n^{-\alpha} \qquad \qquad (\text{eq. 2})$$

where A_n is the area of feature of rank n, A_1 is the area of the biggest feature (rank 1), and α is the characteristic exponent that characterizes the decay of the power law.

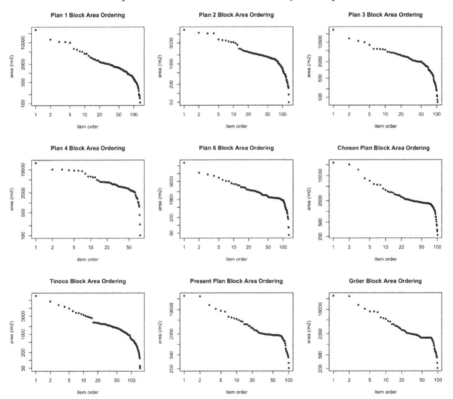

Fig. 1. Block area ordering for the several plans shows a power law dependency throughout most of the ordering range

Block area ordering

The urban block implemented area was the first feature studied. We observed a power law distribution with an exponential decay α ≅ 0.6.

Fig. 1 shows that the area of the blocks of the city decays according to a power law in several plans, although not all the blocks decay in the same way. This behaviour is verified up to blocks of rank 60-80 in the different orderings when, due to the lack of usable area, the block areas decrease abruptly (cut-off point). These high rank blocks are numerous and tend to occupy the voids left by the existence of big blocks in the city. As can been seen, up to 20% of blocks fall into this cut-off segment. Similarities can also be observed in the behaviour of the ordering between Plans 1 through 3, and in the Tinoco plan, while obvious similarities exist between the Chosen plan and the Present-day plan and between the Chosen plan and the de Gröer plan.

Table 1 shows the values of the characteristic decay for the different block sizes. The values are in the range of 0.50-0.73, a bit smaller than the Zipf law (α = 1), but an admirable consistency exists between them. Zipf first formalized the rank sized rule suggesting that the size of cities' Pr scales should be based on the size of the largest size P1 and its rank r in Pr = P1 / r. Notice that this is the same expression as seen in Eq. 1, with α=1.

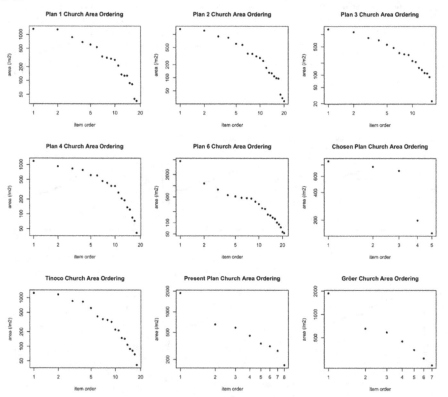

Fig. 2. Church area ordering evidencing the exponential decay of the areas

Church Area Ordering

For the church area, fig. 2 shows that the area rank distribution does not fit a power law, in all cases. This is probably due to the space constraints of the urban fabric. It is observed that in Plans 1, 2, 3, 4, and 5 and the Tinoco plan, the church area clearly approximates an exponential distribution more similar to the cut-off zone. This observation that the church occupied areas decay more quickly than the general block trend is due, most likely, to their special function in the urban fabric, as church occupied areas could support small-sized elements, such as small chapels in the small voids of the city. The present-day church ordering plan and the Gröer plan, with their small number of churches, seem to provide evidence for a closer power law relation rather than an exponential one.

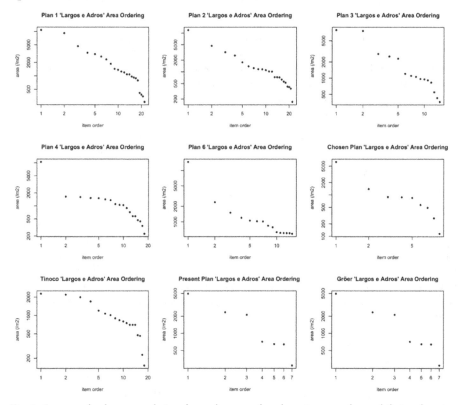

Fig. 3. *Largos* and *adros* area ordering shows the power law decay in some plans, while in others it is closer to an exponential decay in the area

Largos *and* adros *ordering*

Observing the *largos* and *adros* data, we see that no clear single distribution is evident. While in Plan 1 and 2 (and to some point in Plan 3 and the Tinoco plan) the ordering of the areas of *largos* and *adros* is based on the power law, while on the remaining plans we find different behaviours. It is also important to notice that the behaviours of these curves seem not to be coupled with the church areas orderings, although some *largos* and *adros* might be part of churchyards. In general, the area decay seems to indicate that a power law rule governs the distribution of areas as opposed to the church area distribution seen in fig. 2. This can be easily observed in fig. 3, where the decay of the areas seems to more closely resemble a straight line in the log-log plot. In general, this observation suggests that the data presents a great variety of power law exponents, ranging from 0.5 to 1.2. For some series, as can be seen in the case of Plan 1, the law of exponential decay is clearly obeyed, while in other cases, such as the Chosen plan, this behaviour is not shown. This observation makes it difficult to extract a common rule about the *largos and adros* occupancy of the city in these plans and one needs to take a detailed look at each one of them.

Summary Tables

Tables 1-5 show data obtained from the implementation of the five points stated in this paper's methodology section. Several interesting observations can be made about this data.

The total area of the study is, on average, 51 hectares, of which 49 hectares are of land area, while the remaining area is due to the Tagus River (Rio Tejo). The land area is segmented according to Tables 2, 3, 4, and 5. Table 2 provides a synthesis of the land area's constituents. Tables 3 and 4 decompose the public space and the block areas of the city, respectively. Finally, Table 5 shows the primary relationships between the two main squares of the city in each of the nine plans studied.

Table 5 shows, in percentage, the areas of each plan in relation to the present. The last line shows the relationship between the areas of the two squares in each plan.

Discussion of the Summary Tables

Before beginning the discussion, the reader should note that these tables should be read together with the vector images of the plans in the appendices.

As mentioned, the overall area in analysis is approximately 51 hectares. This value represents the average area of the different plans analysed. This is because each of the nine plans in the analysis presents different scales. As a result, many calculations were required so as to determine the distance between specific buildings that did not suffer any adjustment with the 1755 earthquake. The 51 hectares covered the land areas indicated in the plans, such as river areas. This is an important distinction to identify because the coastline is different within the different plans. The Present-day plan and the de Gröer plan present a larger land area because the river area was not included in these plans, as several landfills have been implemented over the course of the time periods investigated.

From Table 1 it is possible to ascertain that the Tinoco plan presents the lowest area of public space (27%) and that it represents about half of the area available in the de Gröer Plan (50%). This is understandable if one perceives the urban morphology in use in each of these two plans. The morphology in the Tinoco plan is tighter and represents narrower streets. The morphology in the de Gröer plan benefit from wider streets and a larger amount of street area (including the stress in the inner areas of the urban blocks) as compared to the Tinoco plan and other plans analysed.

If one excludes the two main squares in the accounting of public space in the de Gröer plan and the Present-day plan (squares Praça da Figueira and Martim Moniz do not exist in the other plans), it is possible to verify that the public space areas, in the plans proposed for Baixa and in the plan area, are similar, varying between 37% and 46%. However, Plan 6 presents the highest amount of public space, supported by the squares area rather than the street area.

	Tinoco	Plan 1	Plan 2	Plan 3	Plan 4	Plan 6	Chosen Plan	de Gröer	Present Plan
α	0.61	0.67	0.61	0.64	0.50	0.73	0.65	0.72	0.67

Table 1. α for the block size distribution for the different plans is similar among plans

	Tinoco	Plan 1	Plan 2	Plan 3	Plan 4	Plan 6	Chosen Plan	de Gröer	Present
Land Area /ha	**47.7**	**46.4**	**49.8**	**47.8**	**48.3**	**47.6**	**43.8**	**55.2**	**55.2**
Public Space	27%	37%	40%	41%	41%	46%	40%	50%	47%
Block Space	62%	54%	49%	53%	59%	50%	59%	48%	51%
Residual Voids	11%	9%	11%	5%	0%	4%	1%	2%	2%
	100%	100%	100%	100%	100%	100%	100%	100%	100%

Table 2. Distribution of the land area by the elements of urban form

	Tinoco	Plan 1	Plan 2	Plan 3	Plan 4	Plan 6	Chosen Plan	de Gröer*	Present*
Land Area /ha	47.7	46.4	49.8	47.8	48.3	47.6	43.8	55.2	55.2
Public Space (a+b)	**27%**	**37%**	**40%**	**41%**	**41%**	**46%**	**40%**	**50% (45%)**	**47% (42%)**
a) Street Area	48%	45%	56%	51%	50%	65%	59%	66% (74%)	64% (72%)
b) Permanence Space Area (b1+b2)	52%	55%	44%	49%	50%	35%	41%	34% (26%)	36% (28%)
b1) Squares	38%	26%	25%	33%	36%	22%	35%	30% (22%)	32% (23%)
b2) 'Largos' e 'Adros'	14%	29%	19%	16%	14%	13%	7%	4% (5%)	4% (5%)

* In parenthesis, the values obtained when the squares of "Figueira" and "Martim Moniz" were excluded

Table 3. Decomposition of the public space in its elements

	Tinoco	Plan 1	Plan 2	Plan 3	Plan 4	Plan 6	Chosen Plan	de Gröer	Present
Land Area /ha	47.7	46.4	49.8	47.8	48.3	47.6	43.8	55.2	55.2
Block Area (a+b)	**62%**	**54%**	**49%**	**53%**	**59%**	**50%**	**59%**	**48%**	**51%**
a) Residential/Others	79%	79%	78%	82%	83%	84%	73%	91%	85%
b) Notable Buildings (b1+b2)	21%	21%	22%	18%	17%	16%	27%	9%	15%
n. Notable Buildings	24	25	26	22	25	26	13	10	13
b1) Civil Buildings	18%	18%	18%	15%	14%	11%	26%	7%	14%
n. Civil Buildings	6	7	6	5	7	5	8	3	5
b2) Church Area	2%	3%	3%	3%	3%	5%	1%	2%	2%
n. Churches	18	18	20	17	18	21	5	7	8

Table 4. Decomposition of the block areas in its elements

	Tinoco	Plan 1	Plan 2	Plan 3	Plan 4	Plan 6	Chosen Plan	de Gröer	Present
Ratios									
Rossio/Present	82%	79%	84%	80%	109%	81%	99%	100%	100%
Terreiro do Paço/Present	114%	99%	117%	171%	83%	89%	105%	100%	100%
Terreiro do Paço/Rossio	2.2	1.9	2.2	3.3	1.2	1.7	1.7	1.6	1.6

Table 5. Analysis of the main squares of the different plans

From the nine analysed plans, it is possible to verify that seven of the plans present public space area that occupies between the 40% and 50% of the plans. Only Plan 1 and the Tinoco plan approximate Krier's ideal values for public space. Krier says that public space should occupy between 25% and 35% of the plan; however, both Plan 1 and the Tinoco plan deal with a very different morphology than the public spaces proposed by Krier, so the percentages could be assumed to be different. In the case of the analysed plans for Lisbon's Baixa area, the distribution of the permanence areas is not uniform. The biggest areas of public space are concentrated on the edges of the studied area.

Even though the percentage of public space area is higher in the de Gröer plan (50%) than it is in all the other plans, the permanence space in the de Gröer plan is the lowest of the nine analysed plans. This is because during twentieth century, primacy was given to mobility within the public space therefore, justifying its 66% occupancy of the street area.

From Table 1 it is possible to observe that only Plan 4 was able to control the overall space. The lack of residual voids in this plan justifies this situation.

Plans 1, 2, 3, 4, the Chosen plan, and the Tinoco plan all present an identical proportion of street area and permanence area. Therefore, Plan 4 should be highlighted due to the fact that its specific area of public space occupies 50% of the plan.

Still, regarding public space, one should also call attention to the similarities of the street area among Plan 6, the Present-day plan, and the de Gröer plan. When compared with the Present-day plan, the de Gröer plan proposes a new street structure (running north-south) that greatly approximates the north-south street arrangements seen in Plan 6. Both de Gröer and Plan 6 present 10 streets (north-south) in the central area of the plan.

While evaluating the permanence areas of the different plans, it is possible to observe that the plans that present the lowest area of *largos* and *adros* are also those plans which present the lowest number of churches. These are the Chosen plan, the Present-day plan and the de Gröer plan.

From the analysis of the built area in Plan 3, it is possible to verify, at the ground level, that a great uniformity exists in the use of distribution in all the plans, which included places seen in the 1756 competition, in the Chosen plan, and in the plan that anticipated the earthquake (Tinoco's plan). Considering that the principal uses for these buildings was residential, as requested by the competition, one can assume that the residential area varies between 73% and 84%, and that the area occupied by notable buildings would oscillate between 16% and 27%. The Present-day plan and the de Gröer plan present the lowest areas occupied by notable buildings (15% in the Present-day plan and 9% in the de Gröer plan). Instead, notable buildings present the highest areas in other uses (this includes commerce and services that have successively substituted the former residential uses), 85% and 91% in the Present-day plan and the de Gröer plan, respectively.

From the comparative analysis analysing the differences between the two mains squares (Terreiro do Paço and Rossio) it is possible to verify that Terreiro do Paço square occupies a larger square area than the Rossio square. That should be justified by the successive high hierarchical uses acquired by the Terreiro do Paço square throughout history. It is interesting to note the ratio between these two squares in Plan 3, wherein Terreiro do Paço square is 3,3 times larger than Rossio square.

From this comparative analysis, it is also possible to verify that the two main squares maintain their proportion in Plan 2 and the Tinoco plan. In both those plans, Terreiro do Paço square is 2.2 times larger than Rossio square. Similar results can be seen in Plan 6 and in the Chosen plan, where Terreiro do Paço square is 1.7 times larger than Rossio square.

When one compares the dimensions of the two main squares, Terreiro do Paço and Rossio, for each plan and for the Present-day plan, one can conclude that the Chosen plan is the one that most closely represents the closest dimensional areas of the Present-day plan (99% when comparing the Rossio squares and 105% when comparing the Terreiro do Paço squares)

When one compares the dimensions of the two main squares, Terreiro do Paço and Rossio in relation to Plan 2, the Tinoco Plan, and the Present-day plan, one can verify an identical proportion among all three plans. Indeed, in Plan 2, Rossio square occupies 84% of the plan and Terreiro do Paço occupies 117%. The respective percentages for the Tinoco plan are 82% for Rossio and 114% for Terreiro do Paço.

5 Conclusions

This work evaluated the public space in nine plans for downtown Lisbon spanning several hundred years. The Tinoco Plan was the oldest plan, having been developed during the Renaissance. Six other plans refer to the proposals presented at the 1756-58 competition for the reconstruction of Lisbon in the aftermath of the 1755 earthquake (Plans 1, 2, 3, 4, and 6, and the Chosen plan). Two other plans were analysed and evaluated as well. These are the plan by the architect Étienne de Gröer, from the mid-twentieth century, created for the reformulation of downtown Lisbon, and the Present-day plan that represents the plans for Lisbon in the twenty-first century.

The analysis followed a methodology for the quantification of the public spaces and the built areas presented in each plan. Secondly, the results were evaluated against the urban design of each plan and information relative to the history of each plan was applied to the analysis. The main elements analysed in this study coincide with those used by Conzen in similar studies.

The present methodology of public space evaluation might be useful for future applications in several ways: either for the evaluation of urban proposals, or for the redefinition of existing urban spaces. Nevertheless further comparisons with other evaluation models or urban spaces would benefit the proposed methodology.

From a historical and chronological analyses, this investigation has made it possible to verify of the following: The Renaissance plan presents an urban design proposal based on public space ratio of 27%; The Enlightenment period plans present urban design proposals with a public space ratio that varies between 37% and 46%; finally, the twentieth and twenty-first century plans present a public space ratio of 47% and 50%, respectively.

The comparative quantification of the public space evaluation has allowed us to verify the existence of a great uniformity in the public space areas among all the nine plans analysed. The reason for this rests on the fact that the urban morphology of the nine plans is practically identical. Nevertheless, that same amount of public space areas could be found in an urban design proposal with a distinct urban morphological approach.

From the various plans proposed for the Baixa competition, Plan 6 stands out as offering the highest public space ratio (46%). This is justified by its urban design, which presents the widest streets of all the proposals. Interestingly, the de Gröer Plan, when compared to the Present-day plan, proposes a new street structure (running north-south) that greatly approximates the north-south street urban arrangements presented in Plan 6.

As can be seen throughout all the plans, although Lisbon's main squares, Terreiro do Paço and Rossio, have remained formally distinct, they have maintained approximately the same areas throughout history.

From the area distribution of the different plans studied one can observe that every plan is similar in some ways to all the other plans. This is reflected in the similar power law exponents that were found for the block area distribution in Table 1. The same type of behaviours could not be generalized for other features, such as churches or *largos* and *adros*; however, in some plans power law behaviour was visible, as shown in fig. 2 and 3. This power law behaviour provides evidence that city size features are dependent on each other in a quantifiable and predictable way. This information also complements the analysis results of the urban features shown in Tables 2, 3, and 4 because it demonstrates how a certain percentage of a feature is actually distributed. For example, in the Tinoco plan, the 62% block area is distributed in a power law manner up to approximately rank 30 and then the block size starts declining quickly in an exponential manner. This analysis is then helpful in understanding the broad range of areas associated with each element of the urban fabric.

Based on this evaluation, and supported by quantitative and qualitative parameters of reference (for example, Krier's [1999] theory of the 'good proportion of public space'), one could now redraw each plan. This would be accomplished with the aim of achieving a good proportion of public space and built area. In order to do so, it would just be necessary to add or subtract certain elements.

Therefore, this methodology is useful because it allows a better understanding of the city. It helps the urban planner design new areas of the city (or redesign old ones). It also helps policymakers choose between different alternative plans according to public priorities. Further, it helps the city's citizens understand how certain features of the spaces in which they live are related in certain ways.

Appendices

Vector format showing built areas.	Vector format showing public space	Source (according Portuguese cataloguing norm NP-405-2)
		1. Cidade antes do Terramoto (Tinoco) [Carta topographica da parte//mais arruinada de Lisboa na//forma, em que se achava antes//da sua destruição para sobre ella//se observarem os melhora//mentos necessários//Reducção à escala 1:2500 da planta//que desapareceu no Archivo do Comando//Geral de Engenharia] [Material cartográfico] AUTHOR(S): SILVA, Augusto Vieira da SCALE: Duas escalas gráficas: uma Esc. gráf. de 100 Varas e outra em metros 1/2500 PUBLICATION: [s.l.]: [s.n.], 1898 PHYSICAL DESCRIPTION: 1 map.; ms. a vermelho s. tela; 49x39 cm NOTES: Planta de Lisboa anterior ao terramoto de 1755 redesenhada em 1898 pelo olissipógrafo Augusto Vieira da Silva. REFERENCE (Archive/Library): MC.DES. 1479 ARCHIVE : Museu da Cidade

2. Plano n° 1

[Planta n° 1//Plano da cidade de Lisboa baixa destruída em que vão// sinaladas por linhas de pontinhos de tinta preta as Ruas// traveças, e becos antigos, e sobre o mesmo plano se mostrão// em branco as Ruas melhoradas assim as largas, como as es//treitas de mayor uso, como também sobre os becos, e Ruas me//nores se desenhão novas ruas que se poderão ou escuzar,// ou abraçar ficando os lugares que os edificios occupão la//vados de aguada preta, As Igrejas dos Conventos, Freguesias e Ermidas vão sinaladas com água de Carmim,// e a divizão das Freguesia de cor azul.] [Material cartográfico]

AUTHOR(S): [FONSECA, Pedro Gualter da; CUNHA, Francisco Pinheiro da]
SCALE: Esc. gráf. de 600 Palmos
PUBLICATION: Séc. XVIII (2ª metade)
PHYSICAL DESCRIPTION: Tinta-da-china e aguarela s/ papel; 64,5 x 85,5 cm
NOTES: Exposição permanente do Museu da Cidade
REFERENCE (Archive/Library): MC.DES .975
ARCHIVE : Museu da Cidade

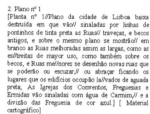

3. Plano n° 2

[Planta n° 2// Planta da Cidade de Lisboa baixa arruinada// em que vão de linhas pretas delgadas as ru//as e travessas antigas, e em branco as ruas de no//vo escolhidas, os edificios novos de carmim claro, // as Igrejas com carmim mais forte, e a cruz, e a // divisão das freguezias de azul.] [Material cartográfico]

AUTHOR(S): [POPPE, Elias Sebastião; POPPE, José Domingos]
SCALE: Esc. gráf. de 1000 Palmos
PUBLICATION: Séc. XVIII (2ª metade)
PHYSICAL DESCRIPTION: Tinta-da-china e aguarela s/ papel; 64,5 x 86,5 cm
NOTES: Exposição permanente do Museu da Cidade
REFERENCE (Archive/Library): MC.DES. 976
ARCHIVE : Museu da Cidade

4. Plano n° 3

[Planta n° 3 // Plano da Cidade de Lisboa baixa des//truida, em que vão signaladas com punctu//ação preta todas as ruas, travessas e becos // antigos, e as ruas novamente escolhidas, e // formadas com toda a liberdade se mostrão // em branco, e os sitios dos edificios novos de // amarello, e as Igrejas e lugares que se con//servão sem mudança de carmim forte, e a // Alfândega do tabaco, Baluarte do terreyro do // Paço e sua cortina, que se devem derribar pa//ra restar formado o grande terreyro do Paço – //vão lavados de huma agoada de carmim, como // também algumas porções de edificios do arco // do açougue té á entrada do Pelourinho, que tão // bem se hão de derribar para complemento do // mesmo terreyro do Paço com semelhante agoada // e a divizão das freg.as com cor azul.] [Material cartográfico]

AUTHOR(S): [CARVALHO, Eugénio dos Santos; ANDREIAS, António Carlos]
SCALE: Esc gráf de 1000 Palmos
PUBLICATION: Séc. XVIII (2ª metade)
PHYSICAL DESCRIPTION: Tinta-da-china e aguarela s/ papel; 64 x110 cm
NOTES: -
REFERENCE (Archive/Library): MC.DES.1782
ARCHIVE : Museu da Cidade

5. Plano nº 4

[Planta nº 4 // Formada ainda com mais // liberdade sem atender a // conservar as Igrejas nos se//us próprios sitios, nem ou/tro algum edificio, como bem // se descobre na delineação do // antigo muyto mais fino.] [Material cartográfico]

AUTHOR(S): [FONSECA, Pedro Gualter]
SCALE: Esc. gráf. de 140 Varas
PUBLICATION: Séc. XVIII (2ª metade)
PHYSICAL DESCRIPTION: Tinta-da-china e aguarela s/ papel; 64 X 84 cm
NOTES: Exposição permanente do Museu da Cidade
REFERENCE (Archive/Library): MC.DES.978
ARCHIVE : Museu da Cidade

6. Plano Escolhido

[Planta topográfica da cidade de Lisboa arruinada também segundo o novo alinhamento dos architectos Eugénio dos Santos Carvalho e Carlos Mardel] [Material cartográfico]

AUTHOR(S): RIBEIRO, João Pedro
SCALE: Esc. gráf. de 2000 Palmos
PUBLICATION: 1947
PHYSICAL DESCRIPTION: Litografia colorida; 5,7cm X 8,3cm
NOTES: Exposição permanente do Museu da Cidade
REFERENCE (Archive/Library): MC.GRA.35
ARCHIVE : Museu da Cidade

[Planta topográfica da Cidade de Lisboa arruinada, // e também segundo o novo Alinhamento dos Architéctos// Eugénio dos Santos Carvalho, e Carlos Mardel] [Material cartográfico]

AUTHOR(S): SILVA, Augusto Vieira da
SCALE: Esc. gráf. de 2000 Palmos
PUBLICATION: Setembro de 1899
PHYSICAL DESCRIPTION: Desenho a tinta da china, aguarelado a rosa e amarelo, 118,9 cm x76,4 cm
NOTES: -
REFERENCE (Archive/Library): -
ARCHIVE : Museu da Cidade

7. Plano nº 6

[Planta // para a renovação // da cidade de Lisboa // baixa destruida ide//ada com toda a li//berdade, assim dë//tro da povoação, co//rmo na marinha së // atender a conserva//ção de couza alguma // antiga, assim sagra//da, como profana.] [Material cartográfico]

AUTHOR(S): [POPPE, Elias Sebastião]
SCALE: -
PUBLICATION: Séc. XVIII (2ª metade)
PHYSICAL DESCRIPTION: Tinta-da-china e aguarela s/papel; 65 X 87 cm
NOTES: -
REFERENCE (Archive/Library): MC.DES.0980
ARCHIVE : Museu da Cidade

8a 8b. Plano do Arquitecto De Groer

[Plano Director de Lisboa, Estudo Pormenorizado da
Transformação da Baixa. Saneamento dos
Quarteirões e Melhoramento da Circulação] [Material
cartográfico]

AUTHOR(S): Etienne De Groer
SCALE: -
PUBLICATION: 27.09.1948
PHYSICAL DESCRIPTION: esquissos, dim : 400
larg. x 500 alt., lápis de cor sobre papel colado em
cartão
NOTES: in TOSTÕES, A. & ROSSA, W. (2009)
Lisboa 1758 : O plano da baixa hoje. IN CML (Ed.)
ROSSA, Walter ed. Lisboa, Câmara Municipal de
Lisboa, Pelouro de Urbanismo e Reabilitação Urbana
e o Pelouro de Cultura, Educação e Juventude.
REFERENCE (Archive/Library): -
ARCHIVE: Colecção Particular

[Plano Director de Lisboa (detalhe)] [Material
cartográfico]

AUTHOR(S): Etienne De Groer
SCALE: -
PUBLICATION: 1948
PHYSICAL DESCRIPTION: esquissos, dim : 42
larg. x 32,5 alt., lápis de cor sobre papel colado em
cartão
NOTES: in TOSTÕES, A. & ROSSA, W. (2009)
Lisboa 1758 : O plano da baixa hoje. IN CML (Ed.)
ROSSA, Walter ed. Lisboa, Câmara Municipal de
Lisboa, Pelouro de Urbanismo e Reabilitação Urbana
e o Pelouro de Cultura, Educação e Juventude.
REFERENCE (Archive/Library): -
ARCHIVE: -

9. Actual

Cartografia Actual da CML

Acknowledgments

We would like to thank architect Sandra Gerardo for the work done in the preparation of
some of the materials contained in this paper and for vectoring the base plans for analysis.

Notes

1. The fact that the Chosen plan is missing is known to almost all the researchers that have
   studied the post-earthquake Lisbon [Sampayo and Rodrigues 2009]. A recent investigation
   regarding this meter [M. Santos 2009] has contributed with a new theory, in which the known
   Plan 6 is considered to be the missing Plan 5.
2. Concepts within the permanence spaces, according to international glossaries of urban form,
   are generally designated as squares. However, in the Portuguese language, these spaces present

different meanings. It is important to stress this issue because some of these spaces affect changes in their designation, in different periods of time, when specific characteristics also change. For example the pre-earthquake Largo dos Remolare was renamed Praça dos Remulares after the earthquake and is known today as Praça Duque da Terceira. Regarding concepts of urban form in Portuguese literature see also [Dias 2000] and [Tudela 1997].

3. However, there are researchers that refer to the concept of square for other periods in history. For example, Françoise Choay refers to the square in the Middle Ages [Choay and Merlin 1988].

4. Besides the archives of "Gabinete de Estudos Arqueológicos da Engenharia Militar"(GEAEM) and of the City Museum, other archives were also investigated for the work presented in this paper. These were the archives of "Gabinete de Estudos Olisiponenses" (GEO) and the digital archive of urban cartography (CEURBAN).

References

BATTY, M. 2007. *Cities and Complexity: Understanding Cities with Cellular Automata, Agent-Based Models, and Fractals.* Cambridge, MA: MIT Press.

BENEDIKT, M. L. 1979. To take hold of space: isovists and isovist fields. *Environment and Planning B: Planning and Design* **6**: 47-65.

BLANCHARD, P. and D. VOLCHENKOV. 2008a. Exploring Urban Environments By Random Walks. In vol. 1021, pp. 183-203 of *Stochastic and Quantum Dynamics of Biomolecular Systems: Proceedings of the 5th Jagna International Workshop.* DOI 10.1063/1.2956796.

――――. 2008b. *Mathematical Analysis of Urban Spatial Networks.* Berlin, Heidelberg: Springer.

CARMONA, M., S. TIESDELL, T. HEATH and T. OC. 2010. *Public Places – Urban Spaces: The Dimensions of Urban Design.* 2nd ed. Oxford: Architectural Press.

CASTEX, J., J. C. DEPAULE. and P. PANERAI. 1977. *Formes urbaines: de l'ilot a la barre.* Paris: Dunod.

CASTEX, J. and PANERAI, P. 1971. Notes sur la structure de l'espace urbain. *Architecture d'Aujourd'hui* **153**: 30-33.

CHOAY, F. and MERLIN, P. 1988. *Dictionnaire de l'urbanisme et de l'aménagement.* Paris: Presses Universitaires de France.

CONZEN, M. R. G. 1960. *Alnwick, Northumberland: A study in town plan analysis.* London: Institute of British Geographers.

DIAS, F. D. S. 2000. Tipologia de espaços urbanos: análise toponímica. Pp. 301-310 in *III Jornadas sobre toponímia de Lisboa,* António Trindade, Paula Machado, Teresa Sancha Pereira, coord. Lisbon: CM.

EULER, L. 1741. Solvtio Problematis Ad Geometriam Sitvs Pertinentis (E53). *Commentarii academiae scientiarum Petropolitanae* **8**: 128-140. Rpt. *Leonhard Euler Opera Omnia,* series 1, vol. 7, pp. 1-10.

FRANÇA, J.-A. 1987. *Lisboa Pombalina e o Iluminismo.* Lisbon: Bertrand Editora.

HEITOR, T. K. M., J. MUCHAGATO, and A. TOSTÕES. 1999. Breaking of the medieval space. The Emergence of a New City of Enlightenment. In: *Second International Space Syntax Symposium,* vol. II. Brasilia.

HILLIER, B. and HANSON, J. 1989. *The Social Logic of Space.* Cambridge: Cambridge University Press.

KRIER, L. 1978. Fourth Lesson: Analysis and project of traditional urban block. *Lotus International* **19** (*L'isolato urbano / The urban block*): 42-54.

――――. 1979. Communities versus zones: A plan for Sodermalm in Stockholm. *Lotus International* **36**: 22-24.

――――. 1999. *Arquitectura : escolha ou fatalidade.* Lisbon: Estar Editora.

KRIER, R. 1979. Typological and Morphological elements of the concept of urban space. *Architectural Design* **49**: 1-17.

KROPF, K. 1998. Facing up to evolution. *Urban Morphology* **2**: 45-47.

KRÜGER, M. 1998. *A Sintaxe da Cidade de Lisboa. Contribuições Para o Desenvolvimento da Cidade.* Coimbra: FCTUC.

LOPES DOS SANTOS, V. M. V. 1994. O sistema construtivo pombalino em Lisboa em edifícios urbanos agrupados de habitação colectiva. Estudo de um legado humanista da segunda metade do Século XVIII. Contributos para uma abordagem na área da recuperação e restauro arquitectónico do património construído. Ph.D. thesis, Universidade Técnica de Lisboa.

LYNCH, K. 1981. *A Theory of Good City Form*. Cambridge, MA: MIT Press.

———. 1996. *The Image of the City*. Cambridge, MA: MIT Press.

MARAT-MENDES, T. 2002. The Sustainable Urban Form: A comparative study in Lisbon, Edinburgh and Barcelona. Ph.D. thesis, University of Nottingham.

MILLER, J. G. 1978. *Living systems*. Niwot, CO: University of Colorado Press.

MONTEIRO, C. 2010. *Escrever Direito por linhas rectas: Legislação e planeamento urbanístico na Baixa de Lisboa (1755-1833)*. Lisbon: AAFDL.

MOREIRA, M. P. P. C. 1993. Conservation of an historic urban centre a study of downtown pombaline Lisbon. Ph.D. thesis, Institute of Advanced Architectural Studies, University of York.

MORRIS, A. E. J. 1994. *History of Urban Form before the Industrial Revolution*. Essex: Longman.

MOUDON, A. V. 1986. *Built for Change. Neighbourhood Architecture in San Francisco*. Cambridge, MA: MIT Press.

———. 1997. Urban Morphology as an emerging interdisciplinary field. *Urban Morphology* 1: 3-10.

MULLIN, J. R. 1992. The reconstruction of Lisbon following the earthquake of 1755: a study of despotic planning. *Planning Perspectives* 7, 7: 157-179.

MURATORI, S. 1959. *Studi per una operante stiria urbana di Venezia*. Rome: Instituto Poligraphico dello Stato.

PAPAGEORGIOU, A. 1971. *Continuity and Change. Preservation in City Planning*. London: Pall Mall Press.

PEREIRA, L. V. 1983. *A forma urbana no planeamento físico*. Lisbon: Laboratório Nacional de Engenharia Civil.

RAPOPORT, A. 1990. *History and Precedents in Environmental Design*. New York: Plenum.

ROSSI, A. 1983. *The Architecture of the City*. Cambridge, MA: MIT Press.

SAMPAYO, M. and RODRIGUES, D. 2009. The Five Plans for the Aftermath of 1755 Lisbon Earthquake: The Interplay of Urban Public Spaces. ISUF Conference 2009 "Urban morphology and urban transformation". Guangzhou, China.

SAMPAYO, M. G. T. D. 2007. Theoretical fundamentals in the construction of the portuguese public squares of the 18th Century. In: ISUF (ed.) Fourteenth International Seminar on Urban Form. Ouro Preto, Minas Gerais, Brazil.

SAMUELS, I. 1993. Working in the park: changes in the form of employment locations. Pp. 113-121 in *Making Better Places: Urban Design Now*, R. Hayward and S. McGlynn, eds. Oxford: Butterwoth Architecture.

SAMUELS, I. 1997. From description to prescription: reflections on the use of a morphological approach in urban design guidance. *Urban Design International* 2: 81-91.

———. 1999. A typomorphological approach to design: the plan for St Gervais. *Urban Design International* 4: 129-141.

SANTOS, M. H. R. D. 2009. Os seis planos da reconstrução. In: *Jornadas sobre A CIDADE POMBALINA: História, Urbanismo e Arquitectura Os 250 Anos do Plano da Baixa, 2008 Lisboa*. Lisbon: Grupo 'Amigos de Lisboa' and Fundação das Casas de Fronteira e Alorna.

SAVAGE, V. and G. WEST. 2006. Biological Scaling and Physiological Time. Biomedical Applications. Pp. 141-164 in *Complex Systems Science in Biomedicine*, T. Deisboeck and J. Yasha Kresh, eds. Heidelberg: Springer.

SIKSNA, A. 1990. A comparative study of Block size and form (in selected New Towns in the history of western civilisation and in selected North American and Australian City Centres). Ph.D. thesis, The University of Queensland.

SOLÁ-MORALES, M. de. 1978. Towards a Definition. *Lotus International* 19 (*L'isolato urbano / The urban block*): 28-36.

TOSTÕES, A. and W. ROSSA, eds. 2008. Lisboa 1758 : O plano da baixa hoje. Exhibition catalogue. (Eng. trans. *Lisboa 1758 - The Baixa Plan Today*). Lisbon: Câmara Municipal de Lisboa.

TUDELA, J. 1977. *As praças e largos de Lisboa: esboço para uma sistematização caracterológica.* Lisbon: Câmara Municipal Lisboa.

WHITEHAND, J. 1981. *The urban landscape: Historical Development and Management. Papers by M. R. G. Conzen.* London: Academic Press.

———. 1990. Townscape management: ideal and reality. Pp. 147-158 in *The Built form of Western Cities. Essays for M.G. Conzen on the occasion of his eightieth birthday,* T. R. Slater, ed. London: Leicester University Press.

———. 2000. From explanation to prescription. *Urban Morphology* 4: 1-2.

———. 2001. British urban morphology: the Conzenian tradition. *Urban Morphology* 5: 103-109.

ZIPF, G. K. 1949. *Human behavior and the principle of least effort.* Cambridge, MA: Addison-Wesley Press.

About the authors

Teresa Marat-Mendes is a Portuguese architect, Professor at the Department of Architecture and Urbanism at the Instituto Universitário de Lisboa (ISCTE-IUL) and Researcher at DINÂMIA-CET Research Centre on Socioeconomic Change, in Portugal. She graduated in Architecture at the Faculty of Architecture of the Technical University of Lisbon, holds a M.Sc. in Land Use Planning and Environmental Planning from the Faculty of Sciences and Technology from the New University of Lisbon and a Ph.D. in Architecture from the University of Nottingham, United Kingdom. Her research interests includes urban morphology, sustainable urban development, urban planning history, ecological urbanism and linking planning theory into practice. She has published several papers in international journals and books in these fields, and has collaborated in international research and academic projects.

Mafalda Teixeira de Sampayo is a Portuguese architect living and working in Lisbon. She has a degree in Architecture from Faculdade de Arquitectura da Universidade Técnica de Lisboa and an M.Sc. in Urban History on the topic of "An Urban Model of Muslim tradition in Portuguese Cities". She has won of several prizes and honours, including EUROPAN 6 with a project for the rehabilitation of 30ha at Setubal. She is presently Assistant Professor at the Lisbon University Institute (ISCTE-IUL) on a leave while researching for her Ph.D. at ISCTE-IUL/FCUL, concerning the topic "Urban Morphology of Public Space in the eighteenth century". Her main research interests include architecture, history, medieval studies and urban studies.

David M.S. Rodrigues was born in Viana do Castelo, Portugal. He has a degree in Chemistry from IST-UTL and a M.Sc. in Complexity Science from ISCTE-IUL/FCUL on the topic of "Community Detection in Email Networks". He is presently doing his Ph.D. research at ISCTE-IUL/FCUL International Doctoral Program in Complexity Sciences on the topic of "Synchronization in Dynamic Network Processes" (http://www.complexsystemsstudies.eu/DavidRodrigues). He is the recipient of a fellowship from ASSYST - Action for the Science of complex SYstems and Socially intelligent icT (www.assystcomplexity.eu) and is a member of the Complex Systems Society (http://cssociety.org). His main research interests include emergence, self-organization, evolutionary systems, network topologies and social networks.

António Nunes Pereira

Rua Antero de Quental, 35, 4º Dtº
1150-041 Lisbon PORTUGAL
anunespereira@gmail.com

Keywords: proportional systems,
Goa, ad quadratum, Renaissance
architecture, sacred architecture,
harmonic proportions,
commensurable dimensions,

Research

Renaissance in Goa: Proportional Systems in Two Churches of the Sixteenth Century

Presented at Nexus 2010: Relationships Between Architecture and Mathematics, Porto, 13-15 June 2010

Abstract. The sixteenth-century Cathedral of Goa and the Jesuit church Bom Jesus reflect mainly European architectural concepts, before local influences appeared in the Portuguese Christian architecture of the following centuries. The research presented here investigated the use of proportional systems. The results show that both the Cathedral and the Bom Jesus have proportions that are usually found in Renaissance architecture of their time, namely, the "ad quadratum" progression and the use of a 4:3 rectangle.

Introduction

Two churches in the former capital of the Portuguese State of India, the Cathedral (1564-1652) and the Jesuit church Bom Jesus (begun 1594, consecrated in 1605), were begun during the sixteenth century and still reflect mainly European architectural concepts, before local influences appeared in the Portuguese Christian architecture of the following centuries. But even if the churches' typology reflected the contemporary religious architecture in Portugal, a unique formal language was developed in Goa. This language showed Portuguese and of course Italian, but also French and Flemish influence through the circulation of etchings, treatises and artists between Europe and the Orient. It became a specific Goan synthesis of the European Renaissance, unknown elsewhere in its purest form [Pereira 2003: 235-239; Pereira 2005: 315-321].

The central question of the following research project was to inquire about the use of proportional systems in these churches' designs and consequently to find out how much they owned to the European culture of the Renaissance. The project's outcome shows that both the Cathedral and the Bom Jesus have proportions that are usually found in Renaissance architecture of their time. In both buildings the proportional systems coincide in ground plan, elevation and sections. This demonstrates that proportions were consciously used to design a whole building, thus reflecting an attempt to achieve harmony through similar dimensions of different parts of a building. But whereas the Cathedral was designed with the *ad quadratum* progression, the proportions of Bom Jesus indicate the use of a 4:3 rectangle. Neither of these churches presents a highly complex proportional system based in musical theory such as Paul von Naredi-Rainer found in Alberti's Tempio Malatestiano and Palazzo Rucellai [Naredi-Rainer 1995: 163-172) or Stefan Fellner in the three Paraguay churches of the Swiss Jesuit Pater Martin Schmid (1694-1772) [Fellner 1993]. Rather, the churches in Goa show a pragmatic use of simple ratios between numbers such as 2:1, 3:2 or 4:3. If music was at all considered in the design process of these churches, their designers chose to privilege the intervals of octave, perfect fifth and perfect fourth and thus the most harmonic of all. The results of this research also confirm Wittkower's assertion, that Renaissance architects thought and created in commensurable dimensions and seldom used irrational proportions. The proportions and geometric patterns of the Cathedral of Goa and of the Jesuit church Bom Jesus show that during the sixteenth century even in the most remote areas of the Portuguese colonial empire the means to achieve architectural harmony remained very close to those of Europe. This fact enlightens us about the processes of transmission of architecture to the new discovered world in the context of the emerging colonial system.

State of the art

Since Rudolph Wittkower's seminal book *Architectural Principles in the Age of Humanism* of 1949 launched the study of architectural proportions in the Renaissance, there has much research in the field of architectural proportion, a great part of which dealing with Renaissance architecture. This and some of the subsequent, more recent publications, were of determining importance for the present research. In his book *Architektur & Harmonie* [Naredi-Rainer 1995], – which has become a standard work in the German speaking world – Paul von Naredi-Rainer deals with number, measure and proportion in the western architecture in addition to period and style. He shows in various buildings the *ad quadratum* and the *ad triangulum* progressions, as well as incommensurable, musical proportions. Karl Freckmann's *Proportion in der Architektur* [Freckmann 1965], Hans Junecke's *Die Wohlbemessene Ordnung* [Junecke 1982], Barbara Böckmann's *Zahl, Mass und Massbeziehung* [Böckmann 2004] also give very informative examples and clues for the present research. Roman architecture was also considered in the present research, even though it was also known to the majority of non-Italian architects and master-builders of the sixteenth century through Renaissance treatises. Mark Wilson Jones' *Principles of Roman Architecture* [Wilson Jones 2000] is one fundamental publication on this topic, even more so as the author also deals with proportion issues. Paul A. Calter's handbook comprehensive display of proportional systems and geometric constructions in *Squaring the Circle* [Calter 1999] was very useful as a guide book as of their appearance throughout History.

Proportional systems and master builders in Goa

Proportional systems of buildings certainly did not depend only on its owner or function, but also very much on the architect. They reflected the designer's design options as well as his cultural context of origin or training. Therefore, master builders – as little as we might know about them – are key figures to unveil the proportional systems issue. During the Renaissance we find on one hand great innovators like Leon Battista Alberti, who developed new architectural concepts, including proportional systems. According to Naredi-Rainer's research, Alberti designed his buildings using sophisticated musical proportions [Naredi-Rainer 1995: 163-172]. Whereas his Tempio Malatestiano shows simple proportions based on the octave (2:1) and the fifth (3:2), the proportions of the Palazzo Rucellai includes many other intervals' ratios, like the fourth (4:3), major third (5:4), minor and major sixth (8:5 and 5:3) and minor and major seventh (9:5 and 15:8). It is likely that we could find such creators in the intellectual humanist circles mainly in Italy or central Europe. This is the case of the Swiss Pater Martin Schmid (1694-1772), a Jesuit musician who went to Paraguay in 1728. His work is known through Stefan Fellner's doctoral dissertation at the Technische Universität Berlin, who studied the musical proportions in the architecture of Schmid's three churches of wood and clay [Fellner 1993].

But on the other hand there were a great number of architects and master-builders who relied on Vitruvius, but certainly also on the medieval Vitruvian tradition. Their concern was to revive Antiquity through building design, approaching what they thought to be Roman architectural concepts. This is implied in the sixteenth-century Portuguese expression to designate Renaissance architecture: *ao (modo) romano*, i.e., "in the Roman (way)". The goal of these architects and master-builders was not to imitate the creations or to adhere to architectural concepts of Italian architects, but to relate to the Roman Empire. In many building tasks the Roman reference was politically important and expressly demanded from the consigner. This was the case in Goa, where the Portuguese were eager to legitimise their news conquests before the other European nations. But antique Rome was not assessable to all sixteenth-century architects and master builders. Those living and working outside Italy had to look at the work of Renaissance architects (through drawings and etchings) but also to authors of Renaissance treatises, as many of them did not have the possibility to travel to Rome and observe the ruins of Classical Antique. In particular, the books of Sebastiano Serlio were a source of graphic information, easily applicable to the design and building tasks these architects had to respond [Kruft 1995: 80]. The adoption of proportional systems remained close to the pragmatic Roman design approach.

Such a pragmatic approach must have been typical of the most important architects in the Portuguese colonies during the sixteenth century, the so-called king's master builders. The office of the king's master builder enjoyed a position of vital importance in the colonies. It was given to an expert trained above all in military architecture, as he had to design, restore and modernize the fortresses that guaranteed Portugal's military dominance in the Orient. These officers were also able to supervise civil and religious construction, as their education comprehended exercises in civil architecture and the knowledge of the most important architectural treatises, as well as of mathematics [Moreira 1993]. Because the Portuguese humanist Francisco de Holanda travelled to Italy and was offered either the *Libro IV* or the *Libro III* by Serlio himself in 1540 [Deswarte 1981, 252-254] his works became very popular in Portugal. Several features in the very churches which proportions we will analyze later testify that this popularity extended to Goa [Pereira 2005: 175-180, 197-201, 203-207, 305-311]. Vitruvius's *Ten Books* were never completely forgotten and circulated now in translations with prints, of which still exist copies in Portuguese libraries of Jean Martin's French translation and Daniele Barbaro's translation and commentary. The Spanish Daniel de Sagredo also contributed with a Vitruvian treatise in 1526, which was printed in Lisbon in 1541 and 1542. And, of course, every new accomplishment in military architecture was immediately introduced in the architectural training. Thus, during the second half of the sixteenth century the leading master-builders and architects, especially those sent to the Orient, were well versed in the latest developments coming from Italy and Central Europe. Following governmental orders in the name of the royal patron (see below), the master builders would draw plans, give advice, or supervise a building site for a religious order. It is highly probable that this was the case of the Goa Cathedral and that its designer was Inofre de Carvalho, the king's master builder at that time [Moreira 1988].

Goa as the capital of the Portuguese State of India

The Portuguese conquered Goa in November of 1510 through the military action of Afonso de Albuquerque. Even though at the time the town was already a very important commercial place, it remained in the first two decades after the conquest similar to the many town-fortresses the Portuguese held along the African and the Asian Atlantic and Indic coasts. But the Christianization of people and territory began immediately after 1510. Before the first parish church was begun in 1514 and dedicated to Saint Catherine of Alexandria, Albuquerque built several votive chapels, thanking God for his help in this important military victory for the Portuguese Crown. In 1543 two other parish churches were founded: Our Lady of Light and Our Lady of Rosary. Two convent churches were also begun in the middle of the century: St. Francis (later Holy Spirit) in 1521, and St. Dominique, in 1550. Of all these churches, only Our Lady of Rosary still stands in the outskirts of Old Goa. All the others were either destroyed or rebuilt. We know therefore very little about the architecture of these early times. The main information comes from written sources, mostly brief building descriptions included in letters, reports, petitions, complaints, and contracts. From these we conclude that with the possible exception of St. Dominique, all these chapels and churches, as Our Lady of Rosary, were still medieval, either Gothic or Manueline Style.

But things were changing both politically and religiously. In 1530 the capital of the State of India (founded in 1505) was transferred from Cochim in Kerala to Goa. In 1533/1534 Goa became a diocese, in 1557 an archdiocese. Goa was now the centre of the small, but numerous Portuguese possessions spread all over the Orient. Goa was also the residence of both the governor or viceroy of India and of the archbishop. It was the only town in the whole Portuguese Empire to have a status very similar to that of Lisbon. State and religion were entangled, as all the territories in the Orient under Portuguese rule were also under the king's religious patronage. The king's obligations extended to the material support for the missionary work of the Catholic Church. This included not only the financial support of parish churches and convents, but also their construction. In this sense, the large majority of churches and convents were royal buildings, just like fortresses or government facilities were.

It is precisely at this point that our two churches diverge. The Jesuit church Bom Jesus is not part of this group. Whereas the Cathedral is the best example of a royal-religious building in Goa, the Bom Jesus Professed House and church was the result of a solely Jesuit initiative and was carried out without royal support – the crown had had already financed the earlier Saint Paul College (built between 1541 and 1576) and with it the king saw his obligations towards the Society in his capital Goa fulfilled. The different proportional systems we are going to find in each church are certainly a consequence of their different status.

In spite of these issues, there is a formal language common to all churches of the sixteenth century. This can be explained not only by of the smallness of Goa's territory, but also by the very architectural activity of master builders. Religious orders seldom had trained architects in their service. Even the Jesuits, who certainly were the best organized of all religious orders, often complained that they did not have learned architects in Goa for all their needs. In most cases, the only solution was to appeal to the king's master builder. It was most certainly this centralized system that gave rise to Goa's homogeneous architectural language, independent of the tradition of any particular order. The resultant visual homogeneity corresponded to the need for identification of an image of this new religion for the local people. It is not surprising then, that these churches, especially their architectonic "face", tended to display a relatively homogeneous configuration. The codification of the architectural rules of the Renaissance style supported the creation of repeatable models, which was furthered by the dissemination of treatises, in particular those of Serlio with its easily adaptable drawings. In Goa, one very particular type became a typical solution for buildings of this time. The Cathedral and the Bom Jesus church represent different stages on the evolution towards this type, which I considered to have been best concretized in the ruined Augustinian church Nossa Senhora da Graça, of 1597 [Pereira 2005: 320-321]. Nevertheless, the proportional systems of both churches differ according to their function and representativeness. We are going to focus on their analysis after giving a short history of these buildings.

The Cathedral

The Cathedral of Goa was built after a royal order to substitute the old church of Saint Catherine, which had been begun in 1514 as the first parish church in Goa. But in 1533/1534 Goa became a bishopric and in 1557 an arch-bishopric, following the restructuring of the Portuguese bishoprics in the 1550s. In 1562 King Sebastião – or rather the regent in his name, as he was just eight years old – ordered a new cathedral to be built. Although the cathedral was begun shortly after 1564, the work was delayed because of lack of funding. It was during the second decade of the seventeenth century that the building progressed swiftly, and was finished between 1651 and 1652. In 1776 the two upper storeys of the north tower collapsed. There is no record of what they looked like. However, as the cathedral is entirely symmetrical and symmetry (in modern, not in a Vitruvian sense) was a very important feature in Renaissance buildings, the façade's drawn reconstruction shows a north tower very similar to the south one. Similarly to most medieval and sixteenth-century cathedrals in Portugal, the Goan Cathedral adopted a Latin cross plan. The church is a false basilica (no clerestory windows between the nave and aisles) with nave, two side aisles and side chapels, transept and choir. The façade shows three bays with portals in the first storey and rectangular windows in the second. The additional central bay linked to its sides through curved walls resolves the height difference between nave and aisles and hides the roof behind it. An entablature between the storeys marks the gallery floor level over the entrance. Portals, windows and niches have detailed, erudite architectural frames, very close to the figures shown in the treatises, especially those of Serlio.

Apart from its scale, the Cathedral is an inconspicuous building, conceived in a somehow discrete Renaissance style (figs. 1-3). Nevertheless, its design is very coherent and homogeneous, which means that in spite of the long building time the original project must have been respected. Although its designer is unknown, there are strong reasons to believe that Inofre de Carvalho was involved in it, as the Cathedral was in fact ordered by the king [Moreira 1988]. Besides some other master builders who worked there, Júlio Simão is known to have had a prominent role in the building during the first decades of the seventeenth century. It is uncertain though, whether he had much influence on the final result or did little more than carry out the initial design.

Architectural proportions in the Cathedral of Goa

As the Cathedral of Goa is a Latin plan church, the starting point of the proportion analysis was its crossing. Firstly, I considered the crossing, including the built elements, i.e., the piers supporting the groin vault (fig. 4). The side of this square was denominated with the letter A.

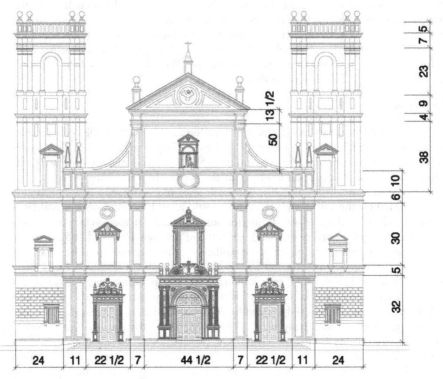

Dimensões em palmo de goa (=0,25⁶ cm)

Fig. 1. Elevation, Cathedral of Goa. Dimensions in goa palms

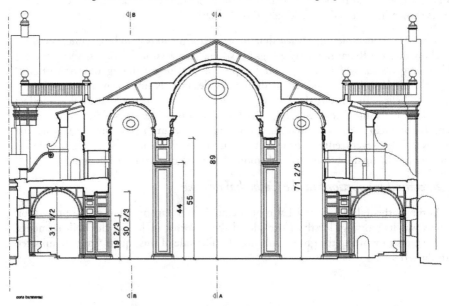

Fig. 2a. Transversal section, Cathedral of Goa. Dimensions in goa palms

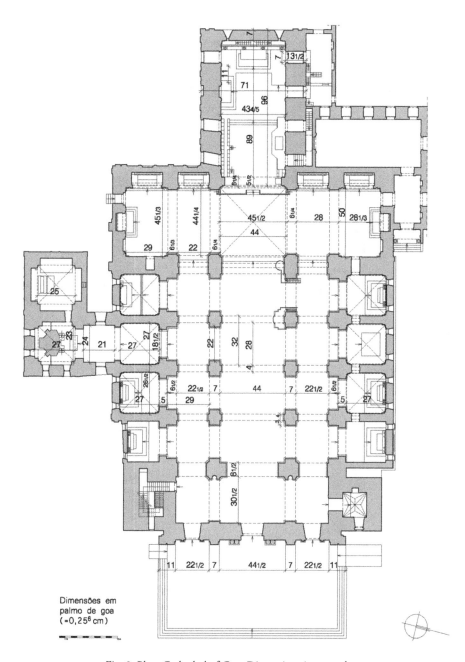

Dimensões em
palmo de goa
(=0,25⁶ cm)

Fig. 3. Plan, Cathedral of Goa. Dimensions in goa palms

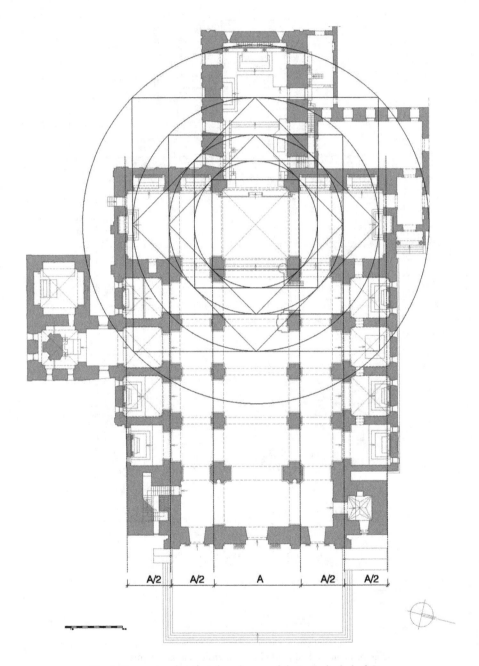

Fig. 4. Proportional analysis based on module A, Cathedral of Goa

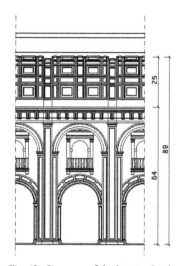

Fig. 2b. Segment of the longitudinal section, Cathedral of Goa. Dimensions in goa palms

Taken from the crossing side A measures ca. 14.86 m (1.61 m for each of the piers and 11.65 m for the span of the crossing),[1] which corresponds roughly to 58 goa palms[2] (14.87 m : 0.256 m = 58.08). It also corresponds to the nave and transept width including the pillars. Setting here the square/circle construction to double the initial square, we get a third square as wide as the side aisles including the pillars forming the outer arches of the side chapels. This proportion corresponds to the architectural morphology, as these pillars have a very similar configuration as the pillars between the nave and aisles.[3] The same construction may be applied to the façade (fig. 5). Even considering only the ground plan, the correspondence between nave and aisles width respectively with the façade central and side bays, as well as between the pillars in the inside and the pilasters – including the angulation – in the outside is evident.

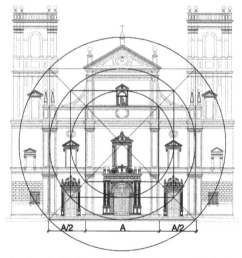

Fig. 5. Proportional analysis of the façade based on module A, Cathedral of Goa

The square A encloses the central bay and the adjoining pilasters, whereas the top square side matches the attic cornice. The third square (side dimension 2A) encloses the inner row of the double corner pilasters, which correspond in the interior to the side chapels' pillars. The outer rows, which do not have any correspondence in the interior, are left out of this construction.[4] The top square line runs along the cornice below the main pediment, where the dove representing the Holy Ghost is. The middle square (side dimension equal to √2A) divides the lateral portals and windows roughly through their middle axis.

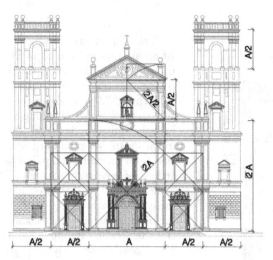

Fig. 6. Proportional analysis of the façade based on module A, Cathedral of Goa

A second drawing (fig. 6) shows further geometrical relations relating to the incommensurable dimension √2A: The height of two storeys of the façade plus attic corresponds roughly to √2A. The height of the additional bay with the sculpture of Saint Catherine triumphing over the Adihl Shah is A/2, whereas the spheres' base over the side acroteria on the pediment corresponds to half of √2A. Further commensurable proportions are the towers width on the façade's basis, which corresponds to half of A, measuring from the inner row of the double corner pilasters; and the height of the towers' last storey, excluding both plinth and balustrade.

The sections (fig. 7) show similar proportions based on the square A. In the transversal section the nave's width equals the height of its side pilasters including the capitals. The entablature above them and the vaulting correspond in height to half of A. From the same point to the roof's top the distance is √2A/2. The aisles' height also matches A, but here the capitals that are at the same level as the walls' cornices are excluded. The longitudinal sections show that the bays between pilasters are A/2 wide.

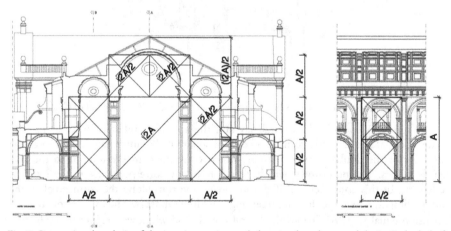

Fig. 7. Proportional analysis of the interior section and elevation based on module A, Cathedral of Goa

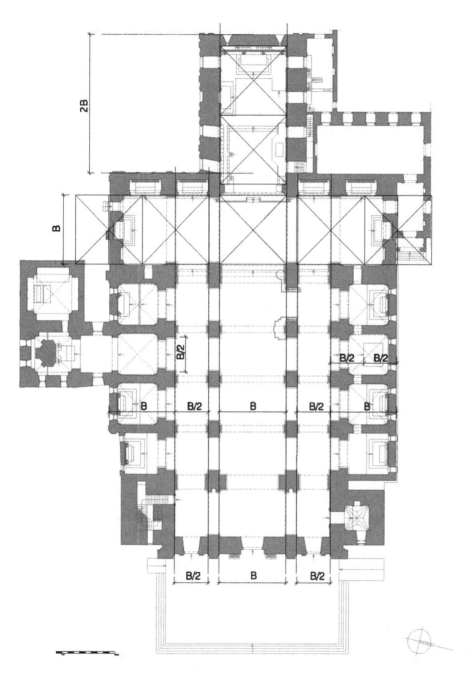

Fig. 8. Proportional analysis based on module B, Cathedral of Goa

So much for proportions built on square A. If we consider only the crossing's free space without angulation, we get another square, which is nominated B (fig. 8). B is ca. 11.23 m, which corresponds to little less than 44 goa palms. The transept is slightly wider (varying between 11.38 m and 11.59 m ≈ 44½ to 45¼ goa palms), but both the choir (11.21m ≈ 43 4/5 goa palms) and the nave (11.29m ≈ 44 goa palms) vary little.

Instead, the transept's arms are $\frac{3}{2}$ B deep, having a *sesquialtera* proportion, the choir is $\frac{2}{1}$ B deep, showing an even more "harmonic" proportion, the *diapason*, as a sacred space. The aisles are half as wide as the nave and the side-chapels including walls and arches are as wide as the nave, corresponding to B.

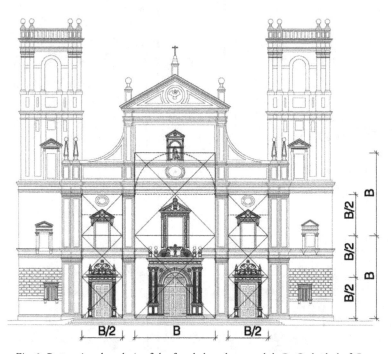

Fig. 9. Proportional analysis of the façade based on module B, Cathedral of Goa

As with the square A, so the façade shows proportions after the square B, although they only concern the bays corresponding to nave and aisles (fig. 9). There is a slight dimensional deviation, as the central bay is 11.39 m and therefore ca. 44½ goa palms wide, in contrast with the crossing's 11.23 m or the nave's 11.29 m (both ≈ 44 goa palms). These deviations were not so evident when we compared the plan with the façade concerning the proportions after square A. Their origin lies mainly in the width variation between pillars in the interior (1.81 m) and pilasters on the façade (1.77 m). This may be due to two reasons: the fact that the Cathedral is plastered and the successive renovations certainly altered the initial shape; but above all that the façade was built almost a century after the foundations and main walls in the interior had been laid, so that minor variations may have occurred. As the crossing is our starting point, we will consider B as corresponding to 11.23 m, even though the nave's width with its 11.29 m is even closer to the considered 44 goa palms. Back to the façade, the central bays – including the entablature that divides them but excluding the upper entablature – have a proportion of *sesquialtera* (3:2), whereas the lateral bays – again excluding the upper entablature – are three times higher than wide, a proportion of *diapason diapente* (3:1). The transversal section (fig. 10) shows that the nave is two times higher (2B) than wide, whereas the longitudinal one evidences that the space between the main pillars also corresponds to B/2.

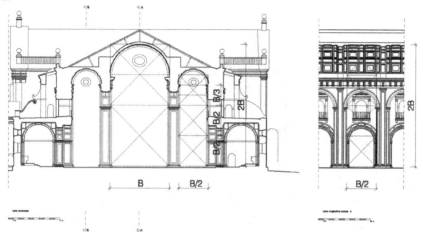

Fig. 10. Proportional analysis of the interior section and elevations based on module A,
Cathedral of Goa

Church of Bom Jesus

Bom Jesus was conceived as the church of the Jesuit Professed House in Old Goa. The Professed House itself was begun eight years before the church, in 1586, as the Jesuits had not the means to finance both building sites simultaneously. The visitor Alessandro Valignano claimed to be the designer of the Professed House. For the church a large rectangular plot by the façade was kept free, as Valignano's drawings sent to Rome show [Wicki and Gomes 1948-1988: XIV, 274-281, 293-295]. Interestingly, the legend in the rectangle representing the church in the drawings sent to Rome by Valignano informed the General of the Society of Jesus that it had not yet been decided, whether the church should have a nave and two aisles or just a nave. Although the architects in Rome had some remarks about the whole Professed House complex – it was considered too much monumental for the Jesuit way, "*modo nostro*" – no comments were made about the vagueness concerning the church's building type [Wicki and Gomes 1948-1988: XIV, 702-706]. Only in 1594 would a legacy of the late captain of Hormuz make it possible to begin it. Bom Jesus was designed with a single nave and two chapel-like spaces to the right and left of the main chapel (figs. 11-13). These spaces and main chapel are narrower and lower than the nave, which means that the architectural space of the Bom Jesus is not cruciform like the Jesuit mother-church Il Gesù in Rome. Rather, the Goan Jesuit church follows the space type of the Portuguese Jesuit churches as Espírito Santo in Évora (1565) and São Roque in Lisbon (1567), as the late art historian Mário Chicó very wisely noticed [Chicó 1956: 267-268]. A cloister-like courtyard was built adjacent to the south side of the church, with cells lining the south side. Two later changes in the church and its surrounding buildings are determining for the proportions analysis: in 1652 a new sacristy was begun to replace an older one; and a few years later the chapel-like spaces by the main chapel were amplified, so that in the south one could receive the tomb of Francis Xavier[5] and the north one the Blessed Sacrament.

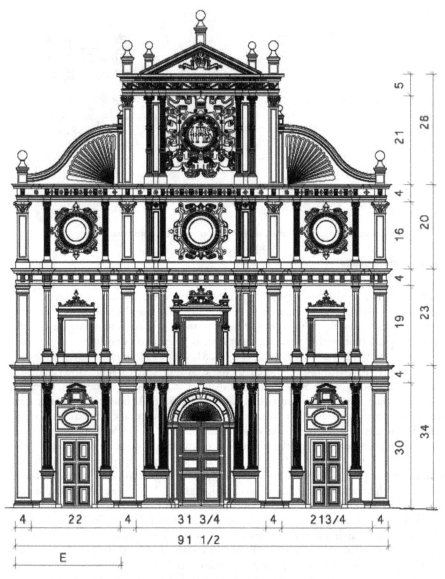

Fig. 11. Façade, Church of Bom Jesus. Dimensions in goa palms

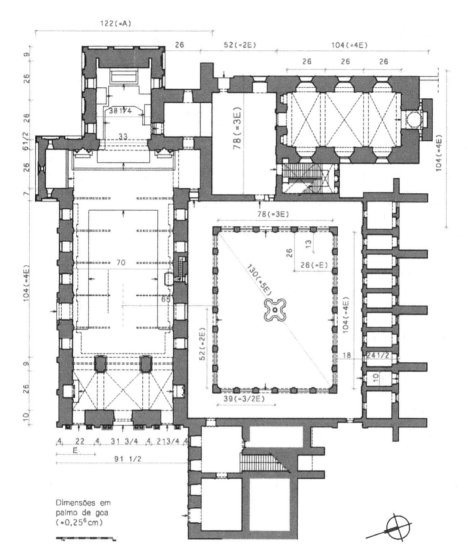

Fig. 12. Plan of the entire complex, Church of Bom Jesus. Dimensions in goa palms

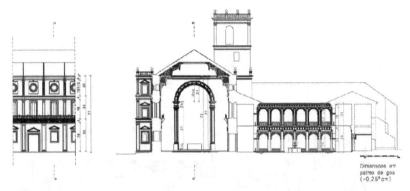

Fig. 13. Interior elevation and section, Church of Bom Jesus. Dimensions in goa palms

Documental evidence indicates that the Bom Jesus's architect was the Jesuit Domingos Fernandes [Wicki and Gomes 1948-1988: XVI, 934], who had most probably already worked with Alessando Valignano in the Professed House [Pereira 2005: 228]. It appears that Domingos Fernandes did not have any special training as an architect, having learned his profession empirically on building sites. Again documental evidence suggests that Fernandes did not design the façade, as it was eventually carried on after 1597 [Wicki and Gomes 1948-1988: XVIII, 808-809]. Its designer, as well as the designers of the seventeenth-century alterations to the Bom Jesus church and sacristy, are unknown.

Architectural proportions in the Church Bom Jesus

As the church Bom Jesus has not a cruciform plan, the approach taken to the proportions analysis was completely different from that of the Cathedral. The starting points were the church's original major width, at the level of the side spaces by the choir, and the form and proportion of the courtyard (fig. 16). Of the two, the church's original major width is no longer evident because of the late 1650s amplifications of the lateral chapel-like spaces by the choir.[6] But the original width is now easily recognizable on the north side, which stands free as it sticks out of the main building. The plaster removal during the 1950s restoration revealed the different stone quality of the seventeenth-century additions, which is more porous and light-coloured than the original laterite stone. This original major width was denominated A and is roughly 122 goa palms.[7] The church's length – excluding the façade's thickness – corresponds to 2A, whereas the choir is A/2 deep. Apparently this simple square construction does not have any further major significance in the plan. But we find it again in the façade (fig. 14). The façade's width corresponds to ¾A and the distances between the portals' axes are ¼A. The ¾A side square inscribing the three main storeys reaches a little above the bases of the spheres' pedestals on both sides of the fourth storey's single bay. A circle inscribing this façade's square – with radius r – touches the upper corner of top pediment. The *ad quadratum* progression shows some interesting results: the three inner squares touch some of the façade's structural elements like pilasters and pedestals, which may or not be coincidences. But it is even more interesting to inscribe two rectangles with the proportion 4:3 in the great façade circle (radius r), one standing and one lying (fig. 15). The standing rectangle encloses the outer pedestals and columns of the ground floor, whereas the horizontal one touches the top of the spheres over the outer façade pilasters.

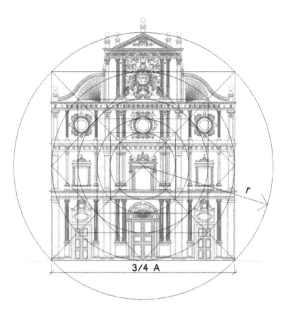

Fig. 14. Proportional analysis of the façade based on module A and radius *r*,
Church of Bom Jesus

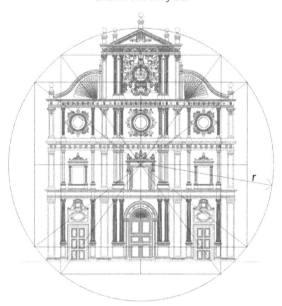

Fig. 15. Proportional analysis of the façade based on radius *r*,
Church of Bom Jesus

But the most astonishing feature is that both coincide in proportion and in
dimension with the inner perimeter of the church's courtyard, as seen in plan. So, unlike
the Cathedral, it is not the *ad quadratum* progression that causes the proportions of
façade and plan to coincide, but a 4:3 rectangle. This courtyard with galleries in two
levels has eight arches in each of the longest and six in each of the shortest sides. If we

consider a module of two arches (module E; fig. 19) we get the 4:3 proportion and a courtyard diagonal corresponding to 5 modules. Half of this diagonal is roughly r – roughly, because in this case dimensions do not completely coincide. Taken from the cloister, r is a commensurable dimension (fig. 17). The arch width – measured from axis to axis of the pilasters – is ca. 3.355 m, i.e., 13 goa palms. A two-arch module is therefore 26 goa palms, which means that the cloister diagonal is 130 goa palms, the half of it corresponding to r (65 goa palms). But taken from the façade r is an incommensurable dimension (fig. 14). If the main circle inscribes a ¾ A square, r is ($\sqrt{2}$ x ¾ A) : 2, i.e., ($\sqrt{2}$ x 91½ goa palms) : 2, which is 64.700265… m, a little less than 64¾ goa palms.

Setting the centre of a circle with radius r in the north-east corner of the Bom Jesus church courtyard inclosing a 4:3 rectangle – equal to the perimeter of the inner courtyard– the rectangle's right side touches the front wall of the staircase and of the sacristy at the southeast corner of the complex (fig. 17).

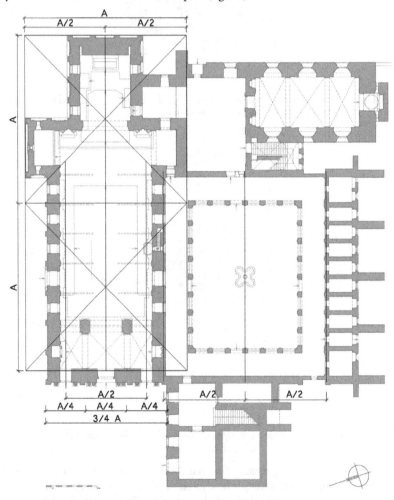

Fig. 16. Proportional analysis of the plan of the complex of the Church of Bom Jesus based on module A

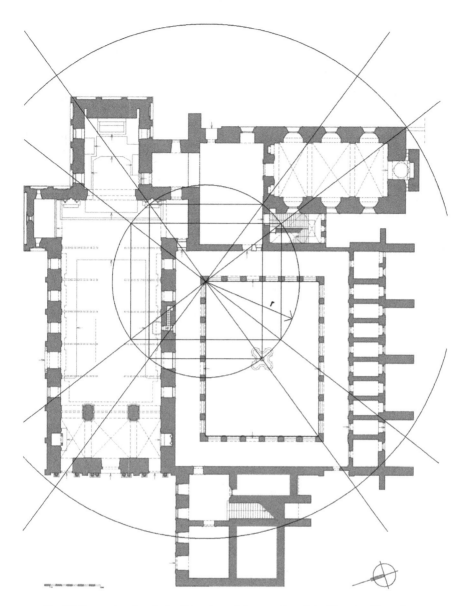

Fig. 17. Proportional analysis of the plan of the complex of the Church of Bom Jesus
based on radius r

The diagonal of a similar, but transversal 4:3 rectangle coincides with the sacristy's
southeast corner. A great circle concentric with the first one and inclosing this point also
touches the corner of the Professed House near its staircase and gets very close to the
façade's northwest corner. But it is in the case of the sacristy that we find the most
interesting coincidences (fig. 18). The very same longitudinal, inner courtyard perimeter
defines the sacristy length. Again, a similar but transversal 4:3 rectangle encloses the
sacristy and the wall section up to the window opening of the anteroom, but excluding
the sacristy's altar volume.

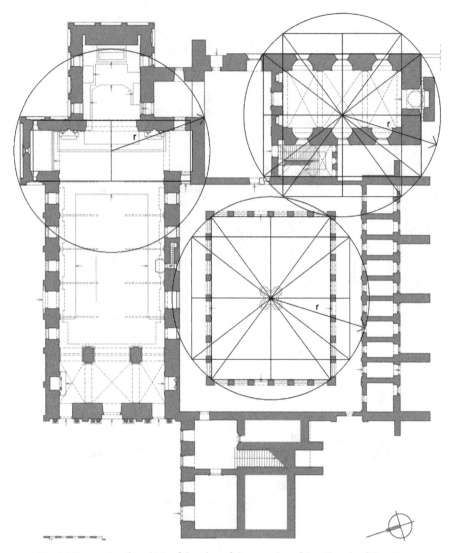

Fig. 18. Proportional analysis of the plan of the complex of the Church of Bom Jesus
based on radius *r*

The similarity between the proportional systems of the church and the sacristy proves
that the church's proportional system of 1594 was still known around the middle of the
following century. It also shows that it was considered important enough to be re-utilized
in the construction of the new sacristy. The following drawings show how module E (the
unit of the 4:3 rectangle) is present in the entire complex of church, courtyard and
sacristy. Spaces like the false transept arms, the choir, the church's nave or the space
underneath the tower (between choir and anteroom to the sacristy) were dimensioned
according to the module E (fig. 19).

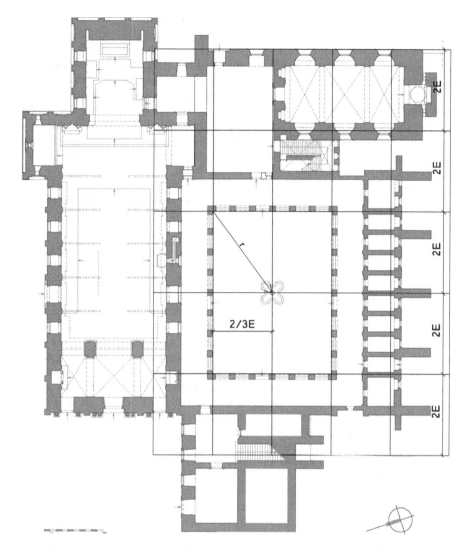

Fig. 19. Proportional analysis of the plan of the complex of the Church of Bom Jesus based on module E

Conclusion: Cathedral versus Bom Jesus

The Cathedral of Goa and the Jesuit church Bom Jesus show two very different proportional systems: the "*ad quadratum*" progression and the 4:3 rectangle. The reason for each respective system must be looked for in the two churches' different status. Nevertheless, the geometric divergence of both systems, the final symbolic significance is the same: the union between Heaven and Earth. In the Cathedral this symbolism has a political overtone, whereas in Bom Jesus it remains mainly religious.

The Goa Cathedral's proportions, mainly based in the Roman *ad quadratum* progression, should be understand as a direct consequence of its double function as a religious and state or royal building. The Cathedral was the principal church in the

Catholic Orient and home of the archbishop and Patriarch of the Orient, but it also was a royal building and therefore an expression of secular, imperial power. The building expresses it most clearly: in the façade's axis there is not only the papal Triple Tiara with Saint Peter's two crossed keys depicted over the main portal, but also the crown of King Sebastião over the central window on the first storey. These two crowns so close to each other in the same building illustrate the entanglement of politics and religion in the process of legitimising the Orient's conquest by Christianity. It is thus not surprising that the chosen proportional system was meant as an intentionally and unequivocal reference to the former Roman Empire the Portuguese were adopting as a model for their colonial expansion in the Orient. With the *ad quadratum* progression the Cathedral's designer achieved the symbolic union between Heaven and Earth as conveyed by a church and simultaneously designed its building form close to the architectural culture of the *romanitas*. Thus, in the Goa Cathedral the ideal of the "Renaissance" as an artistic and political revival of the Roman Empire is materialized to legitimise colonial domination and Christian religious expansion.

In contrast to the Cathedral, the proportional system of Bom Jesus is based on the 4:3 rectangle. Although corresponding to the musical interval of a fourth, I do not believe that this proportion is the result of a "musical" architectural design. The 4:3 rectangle with its diagonal is not only the simplest example of the Pythagorean triangle (3:4:5), but also a rectangle dimensioned on the numbers 3 and 4, whose symbolic meaning in a Catholic missionary context is evident. The number 3 appears for the Divine, not only as a representation of perfection, but also as a symbol of the Holy Trinity [LCI 2004: 1, 524-525]. The number 4 symbolizes the world, referring to the four elements, the four cardinal directions, the four continents[8] and the four seasons of the year[9] [LCI 2004: 4, 459]. Number 4 also has a meaning in the Christian moral, related to the four cardinal virtues, which should conduct the Christian life on Earth – and again the world is symbolised by this concept. The multiplication of 3 by 4 results in 12, a number which has much symbolic significance in many religions and particularly in Christianity [LCI 2004: 4, 582-583]. Its most prominent aspect of interest to Bom Jesus is the meaning of accomplishment of God's Kingdom, which settled on with the announcement of God's word (3) on Earth (4) by the Apostles (12). 12 is therefore a very suitable number for a missionary church as the Bom Jesus. Moreover, we should recall that even before Francis Xavier was canonized in 1622 and his body transferred from old São Paulo to this church in 1624, he was already revered by the Jesuits, who called him the Apostle of India. Those who lived in the Professed House, to which the church Bom Jesus belonged, were following Francis Xavier's steps, that is, they were carrying on the work of the Apostles in a part of the world that the Apostles themselves had never reached (with the exception of Saint Thomas, as was commonly believed amongst the European missionaries). The proportional system of their house church based on the 4:3 rectangle states as much. The missionary aspect is hereby emphasized in opposition to the legitimising of power through the connection between Crown and Church, as it was the case of the Cathedral.

Acknowledgments

The present paper consists in the presentation of research results of a project carried at the Portuguese research centre UNIDCOM / IADE and financed by the Fundação para a Ciência e Tecnologia, Lisbon: "Architecture and Mathematics: proportions systems in two churches of Old Goa of the 16th century."

Notes

1. Like any historical building the Cathedral of Goa is irregular, although the irregularities are very slight for a building of its dimensions. For instances, the crossing's width varies between 11.62 m and 11.67 m, a difference that can be ignored for our purpose.
2. The historical measure unit is the "goa palm", the name of which is not related to the Indian Territory. It is a naval unit of measure of French origin named *goue* that the Portuguese turned into *goa*. It corresponds to 0.256 m [Barata 1965: 155-157; Barata 1989: 191-192].
3. Only the arches' height varies, as those of the side chapels are lower than the central ones, corresponding to the one-storey height of the side chapels.
4. Double pilasters are not a very common feature in Goan Renaissance architecture. They do not appear in Goa before the church of the Carmelite Convent was begun in 1612, which was built by an Italian congregation. Thus, it is very probable that the double corner pilasters at the Cathedral's façade are a later modification of the original project of 1564. Therefore, only the inner row of pilasters is considered in the present analysis.
5. The body of Francis Xavier (1506-1552) was brought two years after his death on Shangchuan Island to São Paulo, the church of the Jesuit College outside Old Goa. After his canonization in 1622 the body was brought to the much more central church of Bom Jesus, where it has rested ever since.
6. This major width is still visible on top of the chapel-like side volumes, as the seventeenth-century amplifications have one fewer storey.
7. Although the same letters are used to nominate geometrical figures in the analysis of both the Cathedral and the Bom Jesus, they are not related with each other.
8. Excluding Australia, which was not officially discovered until 1606, twelve years after the beginning of the Bom Jesus!
9. Ironically, the climate in Goa cannot be divided in four seasons, like those of the European climate. But interestingly enough, in the correspondence of both state clerks and clerics, the terms Summer and Winter are used to distinguish between the Indian dry and rainy seasons.

References

ALBERTI, Leon Battista. 1988. *On the Art of Building in Ten Books.* Joseph Rykwert, Neil Leach and Robert Tavernor, trans. Cambridge, MA: MIT Press.

BARATA, João da Gama Pimentel. 1965. *O «Livro Primeiro da Architectura Naval» de João Baptista Lavanha.* Lisbon: Instituto Português de Arqueologia História e Etnografia.

———. 1989. *Estudos de Arqueologia Naval.* Vol. II. Lisboa: Imprensa Nacional-Casa da Moeda.

BÖCKMANN, Barbara. 2004. *Zahl, Mass und Massbeziehung in Leon Battista Albertis Kirche San Sebastiano zu Mantua.* Hildesheim, Zürich, New York: Georg Olms Verlag.

CALTER, Paul A. 2008. *Squaring the Circle: Geometry in Art and Architecture.* Hoboken: Wiley.

CHICÓ, Mário Tavares. 1956. "Algumas Observações acerca da Arquitectura da Companhia de Jesus no Distrito de Goa". *Garcia de Orta*, número especial, 257-271.

DESWARTE, Sylvie. 1981. "Francisco de Hollanda et les Études Vitruviennes en Italie". In *A Introdução da Arte da Renascença na Península Ibérica.* Coimbra: Epartur. 227-270.

FELLNER, Stefan. 1993. Numerus Sonorus. Musikalische Proportionen und Zahlenästhetik in der Architektur der Jesuitenmissionen Paraguays am Beispiel der Chiquitos-Kirchen des P. Schmid SJ (1694-1772). Diss. Technische Universität Berlin, Berlin.

FRECKMANN, Karl. 1965. *Proportionen in der Architecktur.* München: Verlag Georg D. W. Callwey.

JUNECKE, Hans. 1982. *Die Wohlbemessene Ordnung. Pythagoreische Proportionen in der Historischen Architektur.* Berlin: Verlag der Beeken.

KRUFT, Hanno-Walter. 1995. *Geschichte der Architekturtheorie. Von der Antike bis zur Gegenwart.* 4th ed. Munich: Verlag C. H. Beck.

LEXICON DER CHRISTLICHEN IKONOGRAPHIE (LCI). 2004. Special edition. 8 vols. Rome, Freiburg, Basel, Vienna: Herder.

MOREIRA; Rafael. 1981. "A Arquitectura Militar do Renascimento em Portugal". Pp. 281-305 in *A Introdução da Arte da Renascença na Península Ibérica*. Coimbra: Epartur.

————. 1992. O engenheiro-mór e a circulação das formas no Império Português. Pp. 97-107 in *Portugal e a Flandres. Visões da Europa 1550-1680*. Lisboa: IPPC.

————. 1988. Inofre de Carvalho, a Renaissance Architect in the Gulf. Pp. 85-93 in *Bahrain in the 16th Century. An Impregnable Island*, Monik Kervran, ed. Bahrain.

————. 1993. A Arquitectura Militar. Pp. 137-151 in *História da Arte em Portugal*. Lisbon: Publicações Alfa.

NAREDI-RAINER, Paul von. 1995. *Architektur & Harmonie, Zahl, Maß und Proportion in der abendländischen Baukunst*. 5th ed. Cologne: DuMont.

PEREIRA, António Nunes. 2003. *Die Kirchenbauten in Alt-Goa in der zweiten Hälfte des 16. und in den ersten Jahrzehnten des 17. Jahrhunderts: Zur Entstehung eines Sakralbautypus*. PhD Diss., RWTA Aachen.

————. 2005. *A Arquitectura Religiosa Cristã de Velha Goa. Segunda Metade do Século XVI – Primeiras Décadas do Século XVII*. Orientalia 10. Lisboa: Fundação Oriente.

SERLIO, Sebastiano. 1996. *Sebastiano Serlio on Architecture*, vol. I. Translated from the Italian with an Introduction and Commentary by Vaughan Hart and Peter Hicks. New Haven and London: Yale University Press.

————. 2001. *Sebastiano Serlio on Architecture*, vol. II. Translated from the Italian with an Introduction and Commentary by Vaughan Hart and Peter Hicks. New Haven and London: Yale University Press.

TAVERNOR, Robert. 1998. *On Alberti and the Art of Building*. New Haven, London: Yale University Press.

WICKI, Josef, and John GOMES, eds. 1948-1988. *Documenta Indica*. 18 vols. Roma: Institutum Historicum Societatis Iesu.

WILSON JONES, Mark. 2000. *Principles of Roman Architecture*. New Haven, London: Yale University Press.

About the author

António Nunes Pereira was trained as an architect at the Technical University of Lisbon. He earned his Ph.D. at the Aachen University of Technology, Germany, where he lived for twelve years. His thesis was about the introduction of the European Renaissance in the former Portuguese colony Goa, India, during the sixteenth century, and the development of a specific Goan church architecture. He worked as an assistant to Prof. Dr.-Ing. Hartwig Schmidt at the Department for Conservation of Built Heritage at Aachen University. He also cooperated with several architectural offices in Portugal and Germany. Back in Lisbon since 2003, he has carried on with his research about Renaissance architecture in Goa. With the financial support of the FCT / Lisbon, he has recently concluded a research project about the architectural geometry and proportions of two churches in Goa. In addition to these issues, his research interests include conservation of built heritage, architectural and design theory of the late nineteenth and early twentieth centuries, as well as stage setting in musical dramas. He is currently Associate Professor at the Design University IADE, Lisbon, researcher at UNIDCOM / IADE and, since October 2010, the Director of the Pena Palace in Sintra.

Jong-Jin Park

École Polytechnique Fédérale de
Lausanne (EPFL)
Laboratoire de projet urbain,
territorial et architectural (UTA)
BP 4137, Station 16
CH-1015 Lausanne
SWITZERLAND
jong-jin.park@a3.epfl.ch

Keywords: evolving structure,
urban system, programmatic
moving centre, centroid,
weighted mean, symmetric
optimization, successive
equilibrium

Research

Dynamics of Urban Centre and Concepts of Symmetry: Centroid and Weighted Mean

Presented at Nexus 2010: Relationships Between Architecture and Mathematics, Porto, 13-15 June 2010.

Abstract. The city is a kind of complex system being capable of auto-organization of its programs and adapts a principle of economy in its form generating process. A new concept of dynamic centre in urban system, called "the programmatic moving centre", can be used to represent the successive appearances of various programs based on collective facilities and their potential values. The absolute central point is substituted by a magnetic field composed of several interactions among the collective facilities and also by the changing value of programs through time. The center moves continually into this interactive field. Consequently, we introduce mathematical methods of analysis such as "the centroid" and "the weighted mean" to calculate and visualize the dynamics of the urban centre. These methods heavily depend upon symmetry. We will describe and establish the moving centre from a point of view of symmetric optimization that answers the question of the evolution and successive equilibrium of the city. In order to explain and represent dynamic transformations in urban area, we tested this programmatic moving center in unstable and new urban environments such as agglomeration areas around Lausanne in Switzerland.

Introduction

We consider the urban situation as unstable and the urban form as dynamic. Until today, the analysis of urban morphology was done in a static way, not allowing to follow the evolution nor the successive equilibrium of the city. Beside the general evolutionary form of the city, urban centers are particularly sensitive places of diverse dynamics: they admit more material exchanges or information by concentrating public buildings and places.

We develop and define an interactive and autonomous city system. This enables us to reconstitute the structure of another scale of the city. This "urban system" can develop a new concept of centre[1] by integrating diverse programs.

The concept of the urban system evolves from that of *quartier*, or district. This allows us to re-qualify the evolution and the transformation of the urban centre. Let us define the initial characteristics of the urban system from the point of view of morphogenesis:[2]

- System composed by the various interactive elements

- System as process

- System of the dynamic and mixed centre

The centre transforms continually and displace in an interactive field. The dynamic centre mainly characterizes the evolutionary urban system.

Thus, we focus on three issues about the dynamic centre:

- **Higher geometry complexity** (the complex form of magnetic field by many facilities laid out)

- **Changing value** (the value of programs changes in times following different needs of society)

- **Coordination x, y** (the localization of programs and moving centre)

The *centroid* and the *weighted mean* based on the concept of symmetry are newly proposed to advance these issues in this paper. These mathematical notions integrate into the new urban organization system to identify and represent the dynamic centre.

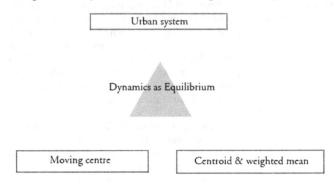

Fig. 1. Three main subjects and dynamics as equilibrium

"Urban system" and "programmatic moving centre"

An urban system of complexity

We suppose that the urban system consists of basic elements, such as habitation or activity groups, public spaces and equipment.[3] The system could be characterized by the appearance of its interactive elements in a perimeter (fig. 2a). In addition, the centre of the system can move in condition (fig. 2b).

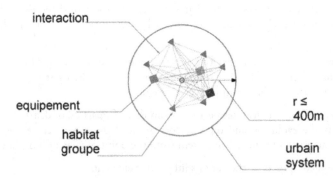

Fig. 2a. Configuration of the urban system. Drawn by the author

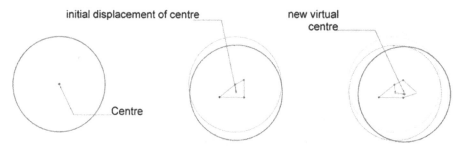

Fig. 2b. The dynamic centre. Drawn by the author

And so, an urban system can be defined by the following formula:

$$US = PS * E * GH * GA = \{PS_{m|m\geq 1},\ E_{N|n\geq 1},\ GH_{p|p\geq 1},\ GA_{p|p\geq 0}\},$$

where n, m and p are the numbers of elements; PS is the subset of public spaces, i.e., plaza, street or park that is often characterized by an important flow of people and that gives public life and activities; E is the subset of equipment, i.e., collective facilities (administration, education (school, university), health (health care centre, hospital), culture (theatre, cinema), religion (church, mosque) etc.); GH is the subset of habitation groups (house, housing); and GA is the subset of activity groups (factory, warehouse, atelier etc.).

In particular, we notice that the proximity becomes a significant measure of the urban system. All the elements influence on each other within a certain distance. The maximum distance from "home" (a habitation group) to a facility in an urban system is of 800 m. Thus, the rayon of interaction of a system is $r=400$ m. It is the general distance allowing the everyday life on foot, or by bicycle in the system. In addition, the appearance of diverse public equipments on the system impact significantly on the dynamics of the centre.

Dynamics of urban centre

We propose a new concept, "programmatic moving centre" which derives from dynamic centre of the urban system based on its evolutionary structure.

This programmatic moving centre also implies the concept of diversity (the phenomenon of growth generates more diverse programs) and of the changing value of programs.[4]

The centre is one of the phenomenological results of the city. It produces and distributes all kind of activities of the city. In fact, the concentration of buildings around public space induces more material exchanges, information, uses and social acts. The monument was often the most prestigious building entity symbolizing the centrality of district, village or city. It has an absolute symbolic value of the system created by and for the man. The absolutely fixed point of the centre by a monument can become dynamic in a field of attraction composed of several groups of the entities. We suppose that complex linkages between several public building entities generate different organization forms so that various types of central configuration can appear. In addition, the changing value of potential programs (according to the need and the change of society) modifies continuously and restructures its centrality.

In fact, the morphology of a city could be understood by its perpetual transformation at the urban system.

To determine the *programmatic moving centre*, several operations should be considered:

- Detection of the potential programs;
- Geometric linkage between these programs;
- Localization of the center by a given geometry;
- Establishment of the urban system in a ray of 400m from the centre of gravity;
- Revaluation of the center according to the structural evolution and changing value of building entities

Our statement is that certain mathematical concepts can be the key words to understand, explain and represent the dynamics based on the *programmatic moving centre* in contemporary cities.

Programmatic moving center and symmetric equilibrium: "Centroid" & "Weighted mean"

Nowadays, the urban area is unstable and it reveals more complex form by the phenomena of growth, sprawl and fragmentation. The urban centre could be considered no more as a static but as a dynamic one! We suggest applying certain mathematic languages such as "centroid" and "weighted mean" to calculate and to represent the dynamic evolution of the urban centre.

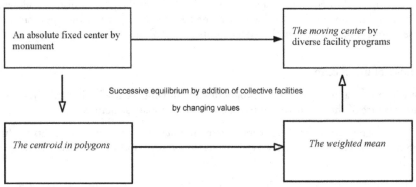

Fig. 3. From an absolute fixed center to the programmatic moving center by integrating "Centroid" and "Weighted mean"

Let us visualize the successive steps of transformation integrating the dynamics of the urban system beyond the static cent*re based on a fixed point by a monument* (fig. 4):

1. Appearance of a system with an historical centre composed of the symbolic building, and habitation groups;
2. Displacement of the centre of the system by adding another equipment;
3. New virtual center among the three equipments and growth of GH (group of habitation);
4. Virtual center reconstituted by different values (*1, 2, 3*) of the equipment;
5. Movement of urban system according to the various centre of gravity.

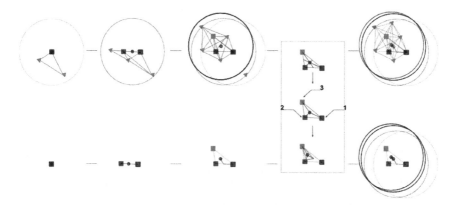

Fig. 4. Dynamics of urban system based on the evolution of gravity, drawn by the author

Legend: △ (in red) - group of habitation, ☐ (in blue) – equipment (facility), ⊙ (in green) - centre of system

In particular, at steps 3, 4 and 5, we start to realize the form of interactive field that can show higher geometry complexity by adding more facility programs. Also, it describes the displacement of point of *centroid* based on the changing value of programs.

Then, let us apply the mathematical formula that makes it possible to obtain the centre of polygonal organization. In the 1st image in the *figure 5* below, the pentagon connecting 5 entities could be broken up into five triangles and each one has its *centroid*. So, the *pentagon centroid* ($\vec{G}$) can be described below:

$$A_1 = \frac{x_1 y_2 - x_2 y_1}{2}$$

$$G_1 = \left\{ \Sigma \frac{(x_1 + x_2)}{3}, \Sigma \frac{(y_1 + y_2)}{3} \right\}$$

$$\vec{G} = \frac{\Sigma A_i \vec{G}_i}{A} = \left\{ \Sigma \frac{(x_i + x_{i+1})(x_i y_{i+1} - x_{i+1} y_i)}{6A}, \Sigma \frac{(y_i + y_{i+1})(x_i y_{i+1} - x_{i+1} y_i)}{6A} \right\}$$

where A is the surface; G is the centre of gravity; and $\vec{G}$ is the centre of gravity in the polygon.

Look at the images in fig. 5, which show the establishment of the centre by the *polygonal centroid*. It shows that the centre moves by applying different values (for example *1,2* and *3*) in pentagonal geometry at the third step.

At the present time, the interaction among the equipments associated with historical buildings in an urban centre, which was represented by a triangle or a simple polygon, will be geometrically much more complex. In addition, the application of the changing value in different time can require more sophisticated calculation. Thus, the *polygonal centroid* above is not sufficient to explain and to represent more complex urban centers.

By considering also the problem of geographic coordination, we could resume the three following problems below:

- Problem of more complex geometries (many facility services laid out);
- Problem of the value of the facilities (values in different times);
- Problem of coordination x, y (localization of programs).

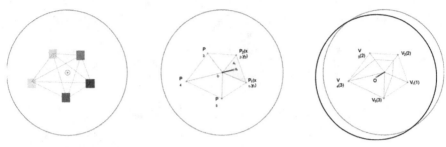

Fig. 5. Establishment of centroid in complex form and displacement of centroid by the changing value of programs in pentagonal organization, drawn the author

Legend: □-facilities, P- coordination of facilities, V- value, O – centre of urban system

The theory of Alain Schärlig for the best localization[5] based on the minimum F (*frais totaux*) helps the understanding and the calculation of the dynamic centre of urban system focused on the facility programs in the research. The concept of the localization theory is based on the equilibrium point that repositions constantly between the components. We also suppose that application of the equilibrium notion is effective to obtain an optimal organization of the dynamic urban system.

Let us propose that the concept of "weighted mean"[6] is appropriate to the determination of *the programmatic moving center* by solving the three problems above through the transformation in time (…, t_{-1}, t_0, t_{+1}, …).

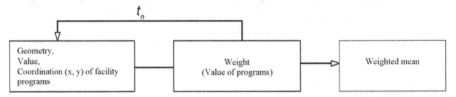

Fig. 6. The problem of programmatic moving centre and weighted mean

We reinterpreted this mathematical law (*the weighted mean*) by integrating it into *the programmatic moving centre*. Then, each facility is considered as a mass that includes geographic coordinates (x, y) with "actual weight" (*the value of the program*).

Here, we determine the application of *weighted mean* to the *programmatic moving center* by weighting the values[7] and by calculating the coordinates (x, y)[8] of the various equipments below:

In Table and graph 7, the value of commercial programs (market and hypermarket) has been amplified (from 1 to 5) like that of school (from 2 to 4). And the appearance of the library and its changing value have made an influence to the dynamic centre. The value of the other programs (church and administration building has been weakened in reverse. In fact, the dynamics of centre (*Point in red→ Point in green→ Point in yellow*) was illustrated by calculating the equipments' positions (x_i,y_i) and those values in different times (t-1→t 0). It showed us the more effective result by applying the *weighted mean*. We will experiment three real cases of the West Lausanne,[9] Switzerland: the center of *Bussigny, Prilly* and *Ecublens* to understand the dynamics of the urban systems in agglomeration area (Table and graphs 8-10).

item no.	name	x position	y position	weight		weight * x	weight * y
1	church	2.5	6	4		10	24
2	commune	2	2	3		6	6
3	school	3	1	2		6	2
4	maket	5	6	1		5	6
5							
	sum	12.5	15	10		27	38
	mean	2.5	3		weighted mean	2.7	3.8

formula = $(w1 * x1 + w2 * x2 + w3 * x3) / (w1+w2+w3)$

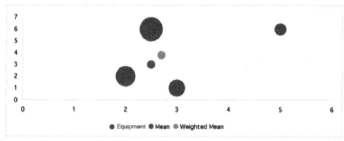

item no.	name	x position	y position	weight		weight * x	weight * y
1	church	2.5	6	1		2.5	6
2	commune	2	2	2		4	4
3	library	5	3	3		15	9
4	school	3	1	4		12	4
5	hypermarket	5	6	5		25	30
	sum	17.5	18	15		58.5	53
	mean	3.5	3.6		weighted mean	3.9	3.53333333

formula = $(w1 * x1 + w2 * x2 + w3 * x3) / (w1+w2+w3)$

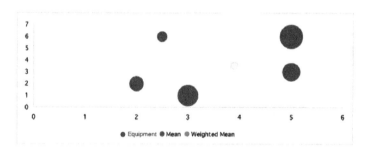

Table and graph 7. t_{-1} to t_0, by various programs and change of value, the moving centre is found on different points as equilibrium

We observe the following:

- The appearance of new equipments (school 2, church 2, market 3 and administration in several directions is less effective to displace the ancient centre.
- The appearance of new equipment (school 2) with its potential value (v=9) can displace easily the centre.
- The appearance of a group of new equipment with potential values on the opposite side of old ones is more effective to displace the centre of the system.

item no.	name	x position	y position		weight		weight * x	weight * y
1	church	1.248	4.853		3		3.744	14.559
2	commune	1.074	4.716		6		6.444	28.296
3	school 1	1.143	4.858		9		10.287	43.722
4	market1	1.243	4.727		1		1.243	4.727
5	market2	1.269	4.701		1		1.269	4.701
6								
7								
8								
9								
	sum	5.977	23.855		20		22.987	96.005
	mean	1.1954	4.771			weighted mean	1.14935	4.80025

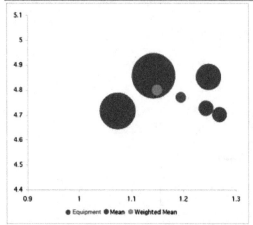

item no.	name	x position	y position		weight		weight * x	weight * y
1	church	1.248	4.853		3		3.744	14.559
2	commune	1.074	4.716		6		6.444	28.296
3	school 1	1.143	4.858		9		10.287	43.722
4	market1	1.243	4.727		1		1.243	4.727
5	market2	1.269	4.701		1		1.269	4.701
6	school 2	1.195	4.784		9		10.755	43.056
7	churchi2	1.031	5.065		3		3.093	15.195
8	marcket3	1.184	4.436		6		7.104	26.616
9	administration	1.003	4.696		3		3.009	14.088
	sum	10.39	42.836		41		46.948	194.96
	mean	1.154444444	4.759555556			weighted mean	1.145073171	4.755121951

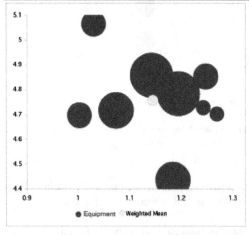

Table and graph 8: t_{-1} to t_0, displacement of the center (1. P in rouge on top → 2. P in green on top →3. P in yellow on bottom), Bussigny

item no.	name	x position	y position	weight		weight * x	weight * y
1	church	5.374	2.942	3		16.122	8.826
2	commune	5.216	3.006	6		31.296	18.036
3	commerce	5.255	2.939	9		47.295	26.451
4	police	5.109	2.937	3		15.327	8.811
5	resturant	5.309	2.854	2		10.618	5.708
6	school	5.178	3.019	9		46.602	27.171
7	nersury	5.219	3.039	9		46.971	27.351
8	marcket1	5.232	2.89	1		5.232	2.89
9						0	0
	sum	41.892	23.626	42		219.463	125.244
	mean	5.2365	2.95325		weighted mean	5.225309524	2.982

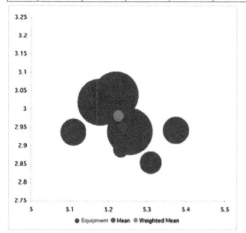

item no.	name	x position	y position	weight		weight * x	weight * y
1	church	5.374	2.942	3		16.122	8.826
2	commune	5.216	3.006	6		31.296	18.036
3	commerce	5.255	2.939	9		47.295	26.451
4	police	5.109	2.937	3		15.327	8.811
5	resturant	5.309	2.854	2		10.618	5.708
6	school	5.178	3.019	9		46.602	27.171
7	nersury	5.219	3.039	9		46.971	27.351
8	marcket1	5.232	2.89	1		5.232	2.89
9	school2	5.309	3.19	9		47.781	28.71
	sum	47.201	26.816	51		267.244	153.954
	mean	5.244555556	2.979555556		weighted mean	5.240078431	3.018705882

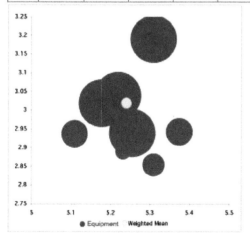

Table and graph 9: t_{-1} to t_0, displacement of the center (1. P in rouge on top→ 2. P in green on top →3. Not in yellow on bottom), Prilly

item no.	name	x position	y position	weight		weight * x	weight * y
1	church	1.849	2.053	9		16.641	18.477
2	commune	1.793	2.032	6		10.758	12.192
3	Restaurant	1.757	1.995	3		5.271	5.985
4							
5							
6							
7							
8							
9							
	sum	5.399	6.08	18		32.67	36.654
	mean	1.799666667	2.026666667		weighted mean	1.815	2.036333333

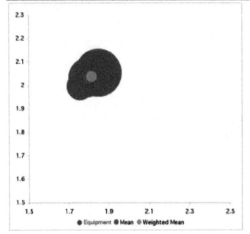

item no.	name	x position	y position	weight		weight * x	weight * y
1	church	1.849	2.053	3		5.547	6.159
2	commune	1.793	2.032	6		10.758	12.192
3	Restaurant	1.757	1.995	3		5.271	5.985
4	administration	1.661	1.761	3		4.983	5.283
5	school	1.77	1.729	9		15.93	15.561
6							
7							
8							
9							
	sum	8.83	9.57	24		42.489	45.18
	mean	1.766	1.914		weighted mean	1.770375	1.8825

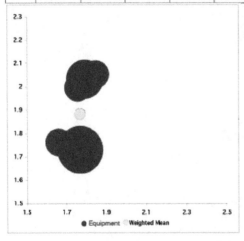

Table and graph 10: t_{-1} to t_0, displacement of the center (1. P in rouge on top→ 2. P in green on top →3. P in green on bottom), Ecublens

Urban morphogenesis representation of West Lausanne

The *urban morphogenesis representation* shows the moving centre in West-Lausanne based on the notion of *centroid* and of *weighted mean*. The magnetic field of urban system has been configured as various geometries from a point to the complex polygon (•→ -- → △→□→★). The numerous equipments and their changing values in times have resulted in urban dynamics. Therefore, the centre moves.

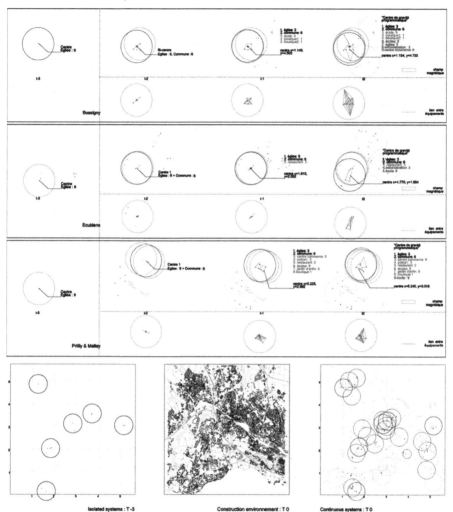

Fig. 11: Evolution of systems and programmatic moving centre: Bussigy, Ecublens and Prilly in West Lausanne, drawn by J-J.Park

Conclusion and discussion

The village or the *quartier* is newly defined as an urban system with its interactive elements: GH, GA, E, and EP as a whole. At this level of organization, the role of the equipments (facilities) revealed through their interaction and relative value is considerable.

The dynamic centre could characterize differently the urban system. The addition, the removal, the displacement of the collective programs and the changing values reinterpret the centrality. Therefore, the city adapts itself to its evolution (according to the needs and the transformation of society).

By developing the concept of the urban centre, it was important to identify and evaluate the existing urban elements and its evolution. Because we supposed that the unstable urban area could be organized more effectively as the indigenous and the new ones became a whole.

Thus, we worked on the *centroid* and the *weighted mean* to identify and evaluate the moving center based on the concept of symmetry. Furthermore, the *weighted mean* was very effective to represent the urban complexity and its centre around dense and mixed buildings. We also expect that these concepts will be useful to determine the localization of the grand equipment at a higher scale.

Three major questions could be asked concerning the dynamics of the urban systems:

- Which centrality is to be proposed for the contemporary city?

- What kind of method is the more useful in order to describe the change of urban state?

- Which kind of operation is appropriate for evaluating the new centrality?

The equilibrium can be explained as the organisms' unceasing act of optimization on growing up. With this process, the concept of symmetry could be applied in this paper in order to understand the contemporary city. It is interesting that the dynamics in urban area can be also understood as the act of unceasing optimal organization. The *centroid* and the *weighted mean*, that relate to the concept of symmetry profanely, allowed us describe and illustrate successive equilibrium of the unstable urban area.

So the dynamics of the centre in the urban system could be recognized by the effect of the equilibrium based on unceasing localization of divers collective programs.

The study can be extended to the two parameters that are decisive for a radical transformation of this equilibrium:

- *Strategic localization of the existing and new programs as collective facilities;*

- *Choice of effective programs according to their changing values.*

The localization of a certain program is already an act of perturbation that can provoke dynamics in the urban area. In addition, the synergy effect by the existing and new collective programs could be more considerable in dense and mixed contemporary city.

The dynamics are mainly influenced by growth and transformation of the facilities in the scale of the system. However, it should be well understood in totality including the dynamics of residence building groups in lower scale and the dynamics of the continuous systems in higher scale.

Otherwise, the automatic integration of evolutionary values to the GIS (Geographical Information System) remains a field to be developed for a forthcoming research about dynamic urban morphology.

Notes

1. In the Western European society, the urban centre (from village to metropolis) has often been established by historical buildings like the church or the town hall built around public places. As time goes by, public facility buildings have been built nearby and newly configured a dynamic urban central area.

2. Morphogenesis is the whole mechanisms explaining the reproducible appearance of structures and controlling their form. It is a fundamental question in all sciences of nature. Professor Patrick Berger (EPFL-IA-UTA) has applied this term to explain urban morphology. We apply this term for the dynamic morphology of contemporary cities. We consider here that the form is deduced by the interaction between the elements of the complex urban system. The pattern develops and become a total form with the process of interactions. The morphology of cities is dynamic while crossing its own transformation process (*urban morphogenesis*). It also could be understood as a result of economic optimization.

3. Normally, the equipment represents the whole of the installations, the networks, the buildings that make it possible to ensure the populations the collective services they need. Let us define the "equipment" as collective building units. These facilities become thus structuring elements of the urban system around the other central elements of the city.

4. It is a reflection on the urban centrality of today from the point of view of complex program. The centrality is found in the middle of various collective housing units including the old prestigious buildings e.g. the church or the town hall. The center, which was imposed by an absolute point of prestigious buildings, becomes the virtual and dynamic point in a magnetic field developed by the several housing units and collective equipments. The complex links between building units are created. This new central configuration can generate different form of fluidity, activities and urban life. Most of all, the urban centre transform by the evolution of the collective programs and their values through the time.

5. $\min F = \sum_{i=1}^{n} w_i r_i \sqrt{(X - X_i)^2 + (y - y_i)^2}$

   $w_i r_i = a_i$

6. One defines "*weighted mean*" as the average of a certain number of balanced values of coefficients. In statistics, considering a set of data, $X = \{x_1, x_2, \cdots, x_n\}$, and the corresponding non-negative weights, $W = \{w_1, w_2, \cdots, w_n\}$, the weighted mean $\bar{x}$ is calculated according to the formulae:

   The weighted mean is the quantity $\bar{x} = \dfrac{\sum_{i=1}^{n} w x_i}{\sum_{i=1}^{n} w_i}$ which means that

   $\bar{x} = \dfrac{w_1 x_1 + w_2 x_2 + w_3 x_3 + \ldots + w_n x_n}{w_1 + w_2 + w_3 + \ldots + + w_n}$

   (extracted from Wikipedia, http://en.wikipedia.org/wiki/Weighted_mean).

7. We abstracted the value of equipment from -3 to 10. The arbitrary evaluation of the value on the paper can be criticized but the changing value of the programs is one of the determinant factors to demonstrate the dynamic tendency of urban center.

8. The coordinates *x*: longitude, *y*: latitude. On the paper, we also abstracted the value of each coordinate (x, y) from 0 to 10 to describe and to explain more effectively.

9. We choose the West-Lausanne as a study area in which the urban dynamics has actually arisen from remarkable urban growth with construction in diverse and considerable transformation near Lausanne city.

References

ALEXANDER, Christopher. 2002-2005. *The nature of order.* 4 vols. Berkeley: Center for Environmental Structure.

ALLAIN, Rémy. 2004. *Morphologie urbaine: Géographie,aménagement et architecture de la ville.* Paris: Armand Colin.

BACRY, Henri. 2000. *La symétrie dans tous ses états.* Paris: Vuibert.

BASSAND, Michel. 2004. *La métropolisation de la Suisse.* Lausanne: PPUR.

BEJAN, Adrian. 2000. *Shape and structure, from engineering to nature.* Cambridge: Cambridge University Press.

BERGER, Patrick and Jean-Pierre NOUHAUD. 2004. F*ormes cachées, La ville.* Lausanne: PPUR.

BOURGINE, P. and A. LENSE., eds. 2006. *Morphogenèse.* Paris: Belin.

DARVAS, György. 2007. *Symmetry.* Basel: Birkhäuser.

DE CHIARA, Joseph, Julius PANERO and Martin ZELNIK. 1984. *Time saver standards for residential development.* New York : McGraw Hill.

FRIEDMAN, Yona. 2008. *L'ordre compliqé.* Paris: Editions De L'eclat.

JACOB, François. 1970. l*a logique du vivant, une histoire de l'hérédité.* Paris: Editions Gallimard.

LABORIT, Henri. 1971. l'*homme et la ville.* Paris: Flammarion.

LECUREUIL, Jacques. 2001. La programmation urbaine: nécessité et enjeux, méthode et application. Paris: Le moniteur.

MARSHALL, Stephen. 2009. *Cities, design & evolution.* NY: Routledge.

MERLEAU-PONTY, Maurice.1945. *Phénoménologie de la perception.* Paris : Gallimard.

MORIN, Edgar.1977. *La méthode: 1. La Nature de la Nature.* Paris: Seuil.

PLAZANET, Corinne, ed. 2008. Morphogenèse de la métropole. Master's thesis, Labo.UTA-INTER-EPFL École Polytechnique Fédérale de Lausanne.

ROSEN, Joe. 2008. *Symmetry rules: How science and nature are founded on symmetry.* Heidelberg: Springer.

SCHÄRLIG, Alain. 1973. *Où construire l'usine ?: la localisation optimale d'une activité industrielle dans la pratique.* Paris: DUNOD.

VERON, Jaques. 2006. *L'urbanisation du monde.* Paris: La découverte.

WEYL, Hermann.1952. *Symmetry.* Princeton: Princeton University Press.

ZWIRN, Hervé P. 2006. *Les systèmes complexs: Mathématiques et biologie.* Paris: Odile Jacob.

About the author

Jong-Jin Park, originally from the Republic of Korea, earned a Master's degree in Art and Architecture under the direction of Prof. Patrick Berger and Prof. Inès Lamunière at the École Polytechnique Fédérale de Lausanne (Thesis and Project: *une promenade vers un Paris insulaire*). He has participated in the architectural and urban design studio of Prof. P. Berger (EPFL) as a teaching assistant and with the thesis project "Morphogenesis" in the laboratory UTA from 2005. He is interested in urban morphogenesis and laws of symmetry to represent dynamic urban morphology. His Ph.D. thesis, "Contemporary City Morphogenesis: Multi-scale Model Experiment" was accepted in April 2010.

Vasco Zara

Université de Bourgogne
Département de Musicologie
36, rue Chabot-Charny
21000 Dijon FRANCE
Vasco.Zara@u-bourgogne.fr

Keywords: René Ouvrard,
Rudolf Wittkower, *Architecture
Harmonique*, *Hypnerotomachia
Poliphili*, seventeenth-century
French architecture, musical
proportions, architectural orders,
rhetorical paradigm

Research

From Quantitative to Qualitative Architecture in the Sixteenth and Seventeenth Centuries: A New Musical Perspective

Presented at Nexus 2010: Relationships Between Architecture and Mathematics, Porto, 13-15 June 2010.

Abstract. In Rudolf Wittkower's influential view, Renaissance musical theory, based on Pythagorean and Platonic proportions, is a paradigm of harmony, order, and spatial organisation in architecture from Alberti to Palladio. However, sources from *Hypnerotomachia Poliphili* (1499) and other sixteenth-century French treatises, through René Ouvrard's *Architecture Harmonique* (1679) seem to show another, undiscovered story. The keys to interpretation include a different philological reading of Vitruvian theory of proportion, as Fra' Giocondo's French lessons show. This is a starting point for a particular sixteenth-century passage between two differents conceptions of architecture – from anthropomorphic to rhetoric, from volumetric to linear, and from quantitative to qualitative – which will find a definitive arrangement in the seventeenth century.

1 Introduction

The relationship between music and architecture can be considered from different disciplinary perspectives, for example from those of acoustics, aesthetics, or history. Here I would like to engage with a simpler and yet more controversial approach: the analogical and the proportional. Let me first make it clear that I am not an historian of architecture, but of music. Furthermore, my discussion is somewhat at variance from that presented in the standard text, Rudolf Wittkower's *Architectural Principles in the Age of Humanism* [Wittkower 1949]. Yet I am not the first to raise criticisms of the work of this Warburg-trained historian since its publication in 1949. A debate about Wittkower's analysis of Palladio's villas began in the late sixties, carried out for example in the pages of the *Bollettino del Centro Internazionale di Studi di Architettura Andrea Palladio*, represented vigorously on one side by Eugenio Battisti [Battisti 1968, 1973] and on the other by Decio Gioseffi, who devoted his corrosive verbal skills to a decisive refutation [Gioseffi 1972, 1978, 1980a-b, 1989]. This debate has been extended by Deborah Howard, Malcom Longair, Branko Mitrovic, Elwin Robison, Lionel March [Howard & Longair 1982; Mitrovic 1990, 1998, 1999, 2001, 2004; Robison 1998-99; March 1998, 2010]. I have had the occasion elsewhere to focus on the positions of each of these historians, and have already attempted to offer a methodological overview and an alternative analytical approach [Zara 2007]. A broad revision of Wittkower's basic general conceptual framework has already been suggested by Henry Millon and Alina Payne [Millon 1972; Payne 1994]. More recently the musicologist Luisa Zanoncelli has raised doubts about the extent of the musical knowledge of one of Wittkower's protagonists, Alberti [Zanoncelli 1999, 2007], doubts that had already been raised, although with different nuances, by historians like Paul von Naredi-Rainer, George Hersey, Pierre Caye, Danilo Samsa, and Angela Pintore [Naredi-Rainer 1977, 1982, 1985, 1994; Hersey 1976; Caye and Choay 2004; Samsa 2003; Pintore 2004]. Nevertheless, as the recent systematic synthesis offered by Peter Vergo demonstrates, Wittkower's book is still the mandatory starting point [Vergo 2005]. Let me then summarize Wittkower's thesis.

The point of departure is the *Descritione* or *Memoriale per condur la fabrica de la Chiesa … Sancti Francisci a Vinea Venetiarum*, written by the Franciscan friar Francesco Zorzi, and completed on 1 April 1535. The story is well known: the doge Andrea Gritti, unhappy with the drawings made by Jacopo Sansovino, official architect of the *Serenissima*, hired Zorzi, who suggested that the proportions of the church should be modified according to the principles of the Pythagorean-Platonic doctrine of sonic numbers. This doctrine, based on the Pythagorean concept of the *tetraktys* and on a philosophic *substratum* provided by the dialogues of Plato, maintains that simple relationships between small numbers mirror the harmony between macrocosm and microcosm. The most basic of these relationships are as follows: 1:2, the *diapason* (in musical terms, the interval of the octave); 2:3, the *diapente* (the interval of the fifth); and 3:4, the *diatesseron* (the fourth). These are not new ideas. Zorzi had already provided an exposition of these doctrines in his *De harmonia mundi totius*, published ten years before; this *summa* of Neoplatonism and Christian Cabala was widely distributed in Latin and in translation, but immediately came under some suspicion of heresy. The *Memoriale*, with its combination of Vitruvian anthropomorphism, Solomonic mysticism and exegesis of the Old and New Testaments, would represent the practical expression of Zorzi's ideas [Vasoli 1998; Campanini 2007]. The document was reviewed by several experts (Titian, Sebastiano Serlio, Fortunio Spira and Sansovino himself, among others), and the construction eventually proceeded through different avenues [Foscari and Tafuri 1983]. For Wittkower, the analogical relationship between the two disciplines is not exhausted by a simple transposition of musical intervals to the spatial dimensions of the building:

> The ratios of musical intervals are regarded as binding, and not the building up of consonant intervals into musical harmonies. Nothing shows better than this that Renaissance artists did not mean to translate music into architecture, but took the consonant intervals of the musical scale as the audible proofs for the beauty of the ratios of the small whole numbers 1:2:3:4 [Wittkower 1962: 102].

The contribution that music brings to architecture is to draw it into the *quadrivium*, thus guaranteeing that it could be considered as a *scientia* rather than simply an *ars mechanica*. The Zorzi *memorandum* is generally considered the first document in which musical ideas are introduced in a very explicit way as point of reference for architectural design. It also functions as a synthesis of a way of thinking that started with Alberti's famous and now hackneyed *dictum*: 'indeed the very same numbers, through which a most delightful harmony is created when borne into the ears, bring it about that the eyes and the mind are filled with a marvellous pleasure' (*hi quidem numeri, per quos fiat cum illa concinnitas auribus gratissima reddatur, iidem ipsi numeri perficiunt, ut oculi animusque voluptate mirifica compleantur*), and which would continue, in Palladio's *Quattro Libri*, to flourish throughout the sixteenth century. For Wittkower, the seventeenth century was thus a period of decadence. It is true that here and there—in

England, Italy and France—strong links with this tradition remained. But according to Wittkower, in France the tradition became 'doctrinal and didactic':

> There was an important French classicist current, the representatives of which kept alive the Platonic conception of numbers in a doctrinal and didactic sense. François Blondel was perhaps the first architect who gave this academic turn to the old Italian ideas on proportion. Almost a whole book of his 'Cours d'architecture', 1675-83, deals with musical proportions in architecture. His approach to the problem is historic and apologetic, for, in contrast to his Renaissance predecessors, he has to prove a case of which many of his contemporaries were ignorant [Wittkower 1962: 126].

According to Wittkower, architects of the seventeenth century began to break the laws of harmonic proportion in architecture; those of the eighteenth century forgot them entirely.

Here I shall present a picture that leads in a different direction. I do not believe that the seventeenth and eighteenth centuries were quite as decadent as Wittkower claimed. On the contrary, I suggest that this period saw the beginnings of systematic thought about the relationships between the disciplines of architecture and music, and a broad dissemination of the results.

2.1 The Architecture Harmonique of René Ouvrard

The first source I shall consider is the *Architecture Harmonique, ou l'Application de la Doctrine des Proportions de la Musique à l'Architecture*, published at Paris in 1679 [Ouvrard 1679]. This is the first book that addresses this topic exclusively, and the first written by a musician and music theorist [Vendrix 1989; Pauwels 2000]. Unfortunately Wittkower – as he freely confessed – was unable to consult this work, for of the two copies extant after World War II, the one in Paris was inaccessible in the *caveaux* of the Bibliothèque Nationale de France (now: Imp. Rés. V. 1886); the second copy (Lyon, Bibliothèque Municipale, Rés. 367374) was presumably too far away to be consulted easily. Wittkower was thus constrained to rely on secondary literature, from which he adopted some incorrect conclusions, especially from the third volume of Angelo Comolli's *Bibliografia storico-critica dell'architettura civile e arti subalterne* (Rome, 1791). For example, he states that the treatise posited the creation of a hypothetical sixth order of architecture, the harmonic order, something that would have important consequences [Wittkower 1962: 126]. Although Françoise Fichet's anthology on French classicism gives a few extracts from Ouvrard, it is not without mistakes of evaluation [Fichet 1979: 175-182]. So, despite Ouvrard's importance, a partial and fundamentally incorrect vision of his work continues to the present day. However, because of its primacy and originality, this treatise has the potential to offer an incomparable historical view of the status of the proportional analogy between the disciplines of architecture and music.

It must be admitted that the author, René Ouvrard, was not a heavyweight in the world of musical literature. His name is hardly found in the historiography of music. But he was not a newcomer either, having been *maître de musique* for sixteen years in the most prestigious musical institution in France, the Sainte Chapelle du Palais de Paris [Brenet 1910; Cohen 1974, 1975; Vendrix 1988]. Ouvrard, who wrote on theology, combinatory mathematics, mnemonic algebra, acoustics, physics, and the theory and history of music, was thus a typical seventeenth-century polymath in the mould of a Marin Mersenne or an Athanasius Kircher. The *Architecture Harmonique* was conceived as a part of his *magnum opus*, *La Musique Rétablie depuis son origine*, an encyclopedic account of the entire science and practice of music and of its literature, but this treatise was never finished, and remained in manuscript [Zara 2008a]. At least three elements of the *Architecture Harmonique*, a work of only thirty pages, deserve to be mentioned: firstly, the object of the treatise, namely the correspondence between music and architecture; secondly, the more general context; thirdly, the received tradition, and a conceptual shift in architecture, as mentioned above.

1. Ouvrard updates the Pythagorean *tetraktys* by reference to the *senario* introduced by Gioseffo Zarlino in his *Istitutioni Harmoniche* (1558). Zarlino had explained that not all the consonances used in modern music are encompassed by the first four numbers of the *tetraktys* (1:2:3:4), but that they extend as far as the first six integers. Beside the consonant intervals, Zarlino acknowledged that the following ratios produce imperfect consonances: 4:5 (major third), 5:6 (minor third), 3:5 (major sixth), and 5:8 (minor sixth) [Zarlino 1558; Mambella 2008]. So far there is nothing extraordinary here: Ouvrard's operation reflects the real need to revise theory in the light of practice. By his time, these latter intervals were widely accepted in theoretical reflection and were ubiquitous in the practice of musical composition. But the epistemological consequences are very much to the point. For to widen the set of acceptable consonances to the first six numbers means that relationships not previously considered consonant suddenly become so. Thus there are more musical numbers than initially expected. In an appendix to his work (*Addition à l'Architecture Harmonique*) Ouvrard repeats sections of Vitruvius' *De Architectura libri decem*, the architectural canon *par excellence*, drawing attention to the musical relationships mentioned in the text. Ouvrard's intention here is to demonstrate that 'toute les proportions qu'il [Vitruvius] a prescrites, sont toutes Harmoniques, quoy qu'il ne leur ait pas donné ce nom, ou que peut-estre il n'en sçeut pas la qualité' ('that all the proportions prescribed by Vitruvius are harmonic, even if he did not describe them as such, or if he did not know that they were such'). [Ouvrard 1679: 16]

2. Ouvrard was the first theorist to define the fundamental proportional relationship not as those exisiting between *diapason*, *diapente* and *diatesseron*, but instead those between *re*, *fa* and *la* (what we would call a d-minor triad). This relation can readily be transposed downward to *ut*, *mi*, *sol* (what we would call a C-major triad), since the sounds that make up the major triad – the terminology is revealing – are more *plaisants* than those that make up the minor triad:

> Ce bastiment a donc des harmonies, 2, 3, 4, 6, 8, 10, 12, 16, 24, 32, qui feroient en Musique ces accords, Ut, Sol, Ut2, Sol3, Ut3, Sol3, Ut4, Sol4, Ut5: Ou plûtost, comme les plus grandes longueurs des tuyaux ou des cordes font les sons les plus bas, en retranchant le nombre 32, on aura dans cet ordre ces proportions, 24, 16, 12, 10, 8, 6, 4, 3, 2, qui feront en Musique ces sons, Ré, La, Ré2, Fa2, La2, Ré3, La3, Ré4, La4 [...] Ce qui fera ces harmonies, Ut, Sol, Ut2, Sol2, Ut3, Mi3, Sol3. Cette harmonie est plus agreable que la precedente à cause de la Tierce Majeure qui est à celle-cy, Ut, Mi, Sol; au lieu de la Mineure qui est en celle-là, Ré, Fa, La [Ouvrard 1679: 10].

Therefore Ouvrard transforms a universal and metaphysical unit of measure – the ancient Greek terminology referred to proportional relationships – by transposing it to a contingent level and thus transforming it into an immanent sonic reality. Now the fundamental elements are not relationships but real, audible sounds.

3. Ouvrard goes one step further and transforms individual buildings – his model is Solomon's temple – into huge megaphone resonators whose foundations vibrate and whose hollow beams, when blown by the wind, resound as a tangible, physical proof of the validity of the correct application of the natural laws of proportion:

> *Et comme ces proportions étoient d'accord avec celles du Temple, & du Sanctuaire, & que Salomon étoit trop sçavant en Musique pour n'avoir pas mis les Trompettes des Prestres, & les divers Instruments des Levites sur le même ton du Bastiment, de l'Autel, & du Bassin; non seulement on entendoit une harmonie parfaite par cette résonnance, mais tout l'Edifice étoit ébranlé & faisoit un bourdonnement & frémissement agreable, tel que celuy qu'on entend dans les voûtes des degrez, lors qu'on sçait prendre leur ton. Nous voyons un exemple sensible de ce pouvoir de résonnance sur les pierres mêmes, dans un pilier en arcade de l'Eglise de Tours, qui tremble à veuë d'œil, & se remuë, dans l'espace de plus de demy-pied, au son d'une certaine cloche, & demeure immobile au son de toutes les autres, quoy que plus proches de luy, & plus grosses que celle qui le fait trembler* [Ouvrard 1679: 13].

In this way Ouvrard explains the phenomenon of sympathy, to which he would return in his unfinished work *La Musique Rétablie* [Zara 2008a]. This example is modelled on the practice, described by Vitruvius in the fifth book of *De Architectura*, of placing vases strategically around a theatre according to Pythagorean proportions to improve their acoustics [Mourjopulous 2003; Godman 2008]. Significantly, this example testifies to Ouvrard's need to anchor the validity of his laws of proportion in empirical reality. Critics of Ouvrard have often dwelt on this example, associating it to the phonurgic machinery of Ouvrard's learned contemporary, the Jesuit Athanasius Kircher [Gozza 2007 (2009)], though they have misunderstood its origins and considered it simply as an example of baroque extravagance, an inferior counterpart to the ascendant rationalism of a Descartes [Picon 1989; Hersey 2000]. However, it was a reading of Ouvrard's description, alongside treatises like the *Génération Sonore* of Jean-Philippe Rameau, that allowed the mid-eighteenth century architect and theorist Charles-Estienne Briseaux (*Traité du Beau essential*, 1752) to constitute the linear correspondence between spatial dimensions and musical intervals. Briseaux's *Division de la Corde sonore* illustrates this point perfectly, in its insistence on a literal translation of the axiom 1:2 = *diapason* = Octave = *Do-do.* Briseaux is thus the first after Ouvrard to make an explicit correspodence between musical proportions and intervals between real sounds. Theorists before Ouvrard spoke of the *diapason*, and never of particular pitches such as *ut–mi–sol*; after Ouvrard, theorists began to speak of definite pitches. This shift may be observed first in Briseaux. [Briseux 1752: 44; Testa 2003]. This does not intend to offer a 'doctrinal and didactic' vision, as Wittkower would have it, of the early Renaissance tradition as represented by Alberti. I do not believe that the Florentine wanted to achieve such a mimesis: he, as Palladio and his followers, based his work on a metaphysical foundation that could legitimate his knowledge and his activities. Even Ouvrard and Briseux looked for theoretical proofs, and in wanting them to be empirical (that is, demonstrable), they organized their thoughts in a systematic way, and with great subtlety; Briseux, for example bases his own demonstrations on Newton's thesis. Their speculation thus becomes objective doctrine, not one based merely on a subjective, 'doctrinal' interpretation. The aberrant 'melodic' interpretations of the architecture of Alberti and Palladio extended later on to every building and every era, as if a series of notes actually issued from the architect's compass, are due to a misunderstanding brought about by removing Ouvrard's work from its context [Bonhôte 1971; Kappraff and McLain 2005]. Once this has happened, the interpretation begins to look dogmatic, since it fails to take into consideration the specific cultural context of production. For this reason I am reluctant to accept analogical interpretations like that offered by George Hersey, who maintained that Ouvrard's work was natural and mimetic. For example, Hersey treated François Blondel's interpretation of Ouvrard's ideas (see below) as normative, applying for example the musical correspondence of the Attic base to Piranesi's candelabrum, designed in 1770, a century after Ouvrard; in Hersey's analysis, this design produces an unresolved dissonant chord (*do–fa–si*). Hersey also applied Blondel's analysis to Bernini's *baldacchino* – designed in 1624-33, fifty years before Blondel – which he claimed to generate a melody in the key of f-minor and in duple meter [Hersey 2000: 22-51]. Explanations of this kind play fast and loose with history, contingent facts, and specific cultural elements [Zara 2005; 2007; 2007 (2009)].

2.2 The Seventeenth-century French context

Did the *Architecture Harmonique* find any contemporary resonances that might permit modern historians to counter Wittkower's well-established picture? At a first glance, the answer seems to be negative. In October 1690, the *Académie d'Architecture* dedicated three sessions to a reading of the treatise, but the conclusion is uncertain:

> L'on a achevé de lire le traité de l'Architecture Harmonique par M. Ouvrard qu'on avoit commencé dans les assemblés précédentes, et, après avoir examiné les exemples qu'il tire de Vitruve avec l'application qu'il en a fait aux nombres harmoniques, on a trouvé qu'à la vérité les distributions s'accordent bien avec ces nombres, mais qu'il y a plusieurs bastiments antiques, de ceux que l'on a estimé, dont la distribution ne s'accorde pas précisément avec cette proportion harmonique [Lemonnier 1912: 203-204].

In July 1692 Christiaan Huygens wrote to Leibniz about this 'petit traité extravagant' [Vendrix 1992]. Ouvrard himself failed to mention this treatise in his testament (dated July 21, 1694), unlike his other works. His friend and correspondent, the abbé Claude Nicaise, in a *Mémoire concernant Maistre René Ouvrard*, written during the same year, likewise failed to mention this work [Zara 2008a]. In 1725, Sebastien de Brossard knew little of the work apart from its title, and the fact that he had lost his own copy of it: 'cette dissertation estoit fort curieuse mais il y a plus de 35 ans que je l'ay perdue et je n'en fait ici mention qu'affin d'insinuer la necessité de la faire chercher' [Zara 2007 (2009)]. Not even Briseux had first-hand knowledge of the *Architecture Harmonique*, and he fails to mention Ouvrard in his *Traité du Beau essentiel*. By contrast, eighteenth-century Italian scholars of architectural theory always cite Ouvrard, but for them the *Architecture Harmonique* represented a kind of Holy Grail: the treatise was conspicuous for its absence, and acquired a kind of *aura imagé*. (In this regard, the case of Wittkower is also emblematic.)

Much more reductively, the context of this work has always been read through the lens of the *querelle* on the nature of architectural proportions carried out between the *médecin-savant* Claude Perrault, translator of Vitruvius and creator of the famous *Colonnade* at the Louvre, and François Blondel, first director of the *Académie d'Architecture* [Herrmann 1973]. For Perrault, such proportions are arbitrary, the result of custom (*accoutumance*) [Scalvini and Villari 1991; Lemerle 2006]. For Blondel, by contrast, they arise from natural laws [Brault 2006; Gerbino 2010]. But this interpretation also requires greater nuance. Elsewhere I have tried to demonstrate how the subject and exposition of the *Architecture Harmonique* reveal a context that exceeded the boundaries of the specific debate over the analogy of musical and architectural proportions, revealing a complex level of speculation that precedes and provides a foundation for the *Querelle des Anciens et des Modernes* [Zara 2006], and perceptibly changes the peculiar point of view on the analogy [Zara 2007 (2009); Zara 2010]. Let me then summarize the principal arguments:

1. Claude Perrault dedicated the first pages of his *Ordonnance des Cinq Especes des Colonnes* to a formal refutation of Ouvrard's treatise. Ouvrard wrote:

> *ce qu'il y a de différence, c'est que les proportions de la Musique consistent tellement dans un point indivisible, que sur le Monochorde l'épaisseur d'un cheveu qui manqueroit à la justesse du son harmonieux se fait sentir; au lieu que la veuë n'est pas si subtile pour apercevoir les petits défauts des proportions* [Ouvrard 1679: 9].

Perrault replied:

> *la connoissance que nous avons par le moyen de l'oreille de ce qui resulte de la proportions de deux cordes dans laquelle l'harmonie consiste, est tout à fait differente de la connoissance que nous avons, par le moyen de l'oeil, de ce qui résulte de la proportions des parties* [Perrault 1683: iv].

This disagreement was carried on before the background of the debate over the natural foundations of music and those of architecture on the other. In his provocative translation of Vitruvius' treatise, Perrault wrote:

> *Les proportions des membres d'Architecture n'ont point une beauté qui ait un fondement tellement positif, qu'il soit de la condition des choses naturelles, et pareil à celuy de la beauté des accords de la Musique, qui plaisent à cause d'une proportion certaine et immuable, qui ne dépend de la fantaisie* [Perrault 1673: 102].

While the detractors of natural law argue that the musical analogy is to be rejected, Perrault's statement witnesses to the fact that this analogy was widely diffused and perceived even as a commonplace.

2. In an attempt to counter the influence of Perrault, François Blondel used Ouvrard's thesis in his *Cours d'Architecture*, a collection of the weekly lessons given to real architects, thus giving it an academic authority and consequently an audience that it could never reach on its own. Blondel also amplified and transformed Ouvrard's ideas, for example by translating the Attic base measures into musical terms [Blondel 1683: 756-760]. This act would have major consequences, unknown to its author; for the musical transposition of the Attic base would take on its own independent life. It was cited and developed in the eighteenth century by Nicolas le Camus de Mézières, the father of *architecture parlante*, in *Le génie de l'architecture* published in 1780 [Le Camus de Mézières 1780: 32] and also by Antonio Vittone and Berardo Galiani. In the nineteenth century it was cited by Baldassarre Orsini and Jean-Michel de Bioul, becoming a regular *topos* in the architectural literature, despite the fact that Ouvrard never mentioned the dimensions of the orders or the bases, or their musical correspondences [Zara 2007

(2009)]. As a result, readers from the eighteenth century to today (as we have noted in the case of Hersey) have generally read Ouvrard by way of Blondel's interpretations [Costantini 2002; Martin 2006]. But the two positions remain quite distinct: Ouvrard tried to demonstrate an analogy, Blondel a mimesis. In any case, Blondel's appropriation again testifies that musical analogy is a familiar element in the construction of the discourse of architectural theory.

3. In the following chapter of his *Cours d'Architecture* [Blondel 1683: 756-760], Blondel attempted to bolster his attempt to provide a workable contrast with Perrault's arguments by introducing the arguments of an architect who was only recently rediscovered by historians: François Bernin de Saint-Hilarion [Scalvini and Villari 1994]. Although his treatise *Des Proportions d'Architecture* was never published and his name was forgotten quite quickly after his death, Bernin was very well known during his life, and he was in contact with Félibien, Le Vau, Mansart, Ouvrard, and all the major architects of his time. Bernin attempted to demonstrate the geometrical nature of architectural proportions. His treatise is not concerned with musical proportions, but, again, this *topos* must be rejected first in order to establish the basis for another. The first line of the manuscript states clearly:

> *S'il est vray que les Proportions Harmoniques sont capables d'accorder les Sons, et que leur Accord produise des effets agréables aux Oreilles comme nous l'expérimantons, il paroist vraisemblable que les Proportions Géométriques doivent aussi avoir la force d'accorder les dimensions, et que c'est principalement de cet Accord que les yeux sont touchez si agréablement à la vüe des beaux Ouvrages de la Nature ou de l'Art* [Bernin de Saint-Hilarion: fol. 2r].

And again, specifying in his fifth chapter (*Quelle sorte de Proportion convient à l'Architecture*) he writes:

> *Puisque les proportions Geometriques ne conviennent point à la Musique: aussi les musicales ne conviennent point à l'Architecture, les unes etant l'Objet des yeux par les diverses dimensions, et les autres etant l'Objet de l'Ouyë par les differens sons; ce qui semble n'avoir rien du tout de commun, tout de mesmes que ce qui touche le Goust ou l'Odorat ne touche pas pour cela la Vüe ou l'Ouyë* [Bernin de Saint-Hilarion: fol. 4v].

4. Three decades before these speculations, Abraham Bosse, teacher of perspective at the *Académie de Peinture*, made drawings in pen-and-ink to accompany the lute compositions of Denis Gaultier, collected in the *Rhétorique des Dieux*. In these drawings, the architectural elements, or more precisely the architectural orders, correspond to the mode of the musical compositions: the Dorian mode corresponds to the Doric order, the Ionian mode to the Ionic order, and so on. The preface to the manuscript is explicit:

> *Et comme chacun de ces modes est propre à exciter certaines passions, et qu'ils sont propres a certaines chants, l'on a representé dans chacun les actions que le mode fait naistre, les Instruments tant anciens que modernes qi luy sont plus convenable, et mesme l'on a observé d'y faire l'Architecture conforme à ces modes* [Buch 1989; Pauwels 2000].

Bosse's mentor, Nicolas Poussin, showed the same attitude, as testified by the famous letter *On the modes*, addressed to his patron Chantelou and dated November 24th 1647

[Jouanny 1911: 370-375; Blunt 1989]. Here Poussin teaches his reader to understand the *ethos* of a pictorial subject through the ancient theory of the musical modes exposed by the ubiquitous Zarlino [Lockspeiser 1967; Pomme de Mirimonde 1972; Mérot 1994; Barker 2000; Stumpfhaus 2005]. A parallel case may perhaps be found in the correspondences between the architectural orders and musical modes outlined in the different choices that Poussin made for the two series of the *Sept Sacrements* [Zara 2008b]. The reference to musical analogy is not confined to Poussin's correspondence. André Felibien, a diplomat who enjoyed an enviable academic career (*conseiller honoraire dell'Académie de Peinture, sécretaire de l'Académie d'Architecture, historiographe des Bâtiments du Roi*) and who in 1676 published the first French architectural dictionary (*Des Principes de l'Architecture, de la Sculpture et de la Peinture*) relates that Poussin's paintings contain an *ethos* that suggests the musical modes: *dorien, lydien, ionien,* and an improbable *lesbien*. Henri Testelin, his successor, did not forget his predecessor's teaching nor, unfortunately, his habitual inaccuracy; according to Testelin, the relevant modes were *phrygien* and *corinthien* (*sic!*) [Allard 1982; Montagu 1992].

5. In his *Parallèle de l'Architecture Antique avec la Moderne*, which Ouvrard read as a kind of moral inspiration, Roland Fréart de Chambray – translator of Palladio and Leonardo, and benefactor of Poussin, to whom he dedicates his *Idée de la Perfection de la Peinture* (1662) – identified the Solomonic order as 'the flower of Architecture and the order by the orders' [Fréart de Chambray 1650: 71]. He did fail to mention the fact that Juan Bautista Villalpando, in his widely received *In Ezechielem Explanationes* [Villalpando 1595-1604] had designed this order, to the most minute details of the metopes and triglyphs, in explicit reliance on the Pythagorean *tetraktys* [Ramirez 1991; Sanchez de Enciso 2008]. Because of the deep musical relationship affirmed in Villalpando's work, Claude Perrault mocked the author, accusing him of mystagogy with his ridiculous 'mystère des proportions' [Hermann 1967].

6. Returning to the *Architecture Harmonique*, we note that these two elements – the order and the musical analogy on a rhetorical and not a solely proportional plan – are present both in the genesis of the work and in its legacy. As we have noted, the treatise arose in reaction to the provocations of Perrault (see above), and as a result of the contest announced by Colbert in 1673 (the year of the opening of the *Académie d'Architecture*) for the creation of a sixth architectural order, the French order, which was to encompass in itself the character of the nation, and which was for this reason to be placed at the top floor of the *cour carrée* of the Louvre as summit and synthesis of all the other orders. At stake was the establishment of a rhetorical paradigm capable of giving life to a precise and peculiar character: the French national order. But Ouvrard was not interested in creating a new order, and merely recalled the musical rules, 'puisque sans la Doctrine des Proportions Harmoniques tous les Ordres d'Architecture ne sont que des amas confus de pierres sans ordre & sans regle' [Ouvrard 1679: 2]. For Ouvrard, a music theorist and pedagogue, proportion means character, as illustrated by the lore of the modes from Pythagoras and Plato and beyond. Perrault recognized the importance of this notion, as is clear from the programmatic 'Préface' of his *Ordonnance des Cinq Especes des Colonnes*: 'il me reste de dire les raisons que j'ai de changer aussi quelque chose dans les Caractères qui distinguent les Ordres; ce qui est une license encore plus grande que celle de toucher aux proportions' [Perrault 1683: xxiij-xxiv]. But unlike Ouvrard, Perrault considered proportion and character as two distinct notions. According to contemporary *Dictionnaire françois* of Pierre Richelet, Perrault considered 'caractère' to means 'la

marque qui distingue une personne, ou une chose d'une autre' [Richelet 1680: 110]. It is a morphological sign, a distinguishing attribute which remains however fundamentally exterior the object, an addition. One century later Le Camus des Mézières criticised Perrault's idea exactly because of this distinction:

> Il paroît qu'on ne peut varier que dans les ornemens & les hauteurs ....
> Rien de plus ingénieux que celui que Perrault a destiné à l'Ordre Français;
> les masses sont à peu près les mêmes que celles de l'Ordre Corinthien.
> Mais les attributs en changent le caractère [...]. L'idée est ingénieuse, mais
> l'ensemble n'est enfin qu'un chapiteau composite, rien de nouveau dans
> les proportions, conséquemment point de sensations qui caractérisent un
> nouvel Ordre [Le Camus de Mézières 1780: 38-39].

This does not mean that Le Camus de Mézières acted in the same way as Ouvrard. It is important to emphasise that sixteenth-century architectural treatises tend to concentrate on the architectural orders, because in the diameter of the orders resides the principle of the rule [Szambien 1986]. This is true for Ouvrard (and this is the reason why he dismissed the creation of a new order in favour on a demonstration of the order of the law), but not for Le Camus de Mézières, who dedicated just a few pages to a discussion of the orders. Like the anthropomorphic paradigm, the notion of order from the universal became subjective during the seventeenth century [Rykwert 1998; Zara 2007 (2009)]. In the new and properly aesthetic value, the musical reference remains, but dismisses its *numerical* basis and now assumes a *rhetorical* model. 'Music means a range, not an essence' [Serravezza 2004: 125], a sentiment illustrated by Le Camus de Mézières:

> Ne pourroit-on pas cependant employer dans cet Ordre des proportions
> mixtes ou participants des deux différens Ordres, comme on emploie les
> demi-tons? C'est une question à resoudre, & sans doute assez délicate [Le
> Camus de Mérières 1780: 38].

Nobody could find a solution. And for the very first time, a dissonance, the semitone, became a positive touchstone. But the presence of these elements, these actors, these words, suggests that such ideas had been circulating for some time. Moreover, the French context makes it clear that the link between music and architecture was now to be established not simply on the basis of numbers or proportions, but more properly on the basis of rhetoric.

3 The rhetorical model and the Hypnerotomachia Poliphili

In the fifteenth century a number of factors gathered independently around the analogy between music and architecture, and served to strengthen the rhetorical model of music and architecture. Although these are more or less independent, they combined to contribute to a redefinition of architecture. The most characteristic factors are as follows: the analogy made by Francesco Colonna in the *Hypnerotomachia Poliphili* (1499) between the 'compositional process' exercised by architects and composers; a misguided emendation to the text of Vitruvius made in 1511 by Frà Giocondo (1433-1515), which made Vitruvius' work seem more deeply dependent on the principle of proportion than is really the case; Colonna's substitution of Vitruvius' terms *ratiocinatio* and *fabrica* with Gaffurio's terms *lineamento* and *prattica*; and finally, the gradual adoption, over the course of the sixteenth century, of the metaphor of an architectural feature, the 'scale' (that is, a ladder or staircase), in the language of music. All these features show independently that the model of rhetoric increasingly complemented the mathematical model of music and architecture current before the sixteenth century.

As we have already noted, the *mystère des proportions* had ancient origins: references can be found, perhaps with different *foci*, in Alberti, Zorzi, Barbaro and Palladio. Philibert Delorme, the first architect in France to write about architecture, announcing a book dealing entirely with the divine proportions – one that was however never to be published – repeats the *exempla* of Zorzi's Venetian *memorandum* about the Venetian church, even giving them in the same order:

> *Nous parlerons des sainctes et divines proportions données de Dieux aux saints pères du vieil testament: comme à son Patriarche Noé, pour fabriquer l'Arche contre le cataclysme et déluge; à Moyse, pour le Tabernacle de l'autel, des tables, des courtines, du parvis et autres; à Salomon, pour le Temple qu'il édifia en Jérusalem ...* [Delorme 1576: fol. 4r].

It is unknown if Delorme had seen the manuscript by the Venetian friar, whether directly in Venice during his stay there (a period that coincided with the construction of the Minorite Church), or indirectly through his teacher Serlio, one of those who had been commission by the doge to examine Zorzi's *memorandum*, and who would spend his last years working in France [Blunt 1958; Pérouse de Montclos 2001]. Nevertheless, it is clear that Delorme's description of the 'divines' contained some musical resonance. As Pérouse de Monclos states, the connexion may be indirect, for Zorzi's *memorandum* was based on the ideas expressed more fully in *De harmonia mundi totius* (see above).

But there is another book that preceded that of Zorzi and, as Carlo Dionisotti suggested, may have inspired it in several respects [Dionisotti 1968]: Francesco Colonna's *Hypnerotomachia Poliphili*, printed and published by Aldo Manuzio in Venice in 1499. This book was born in the same Venetian friary in which Francesco Zorzi lived, and was well known even into the seventeenth century; Félibien describes it as 'a book that all people know today' [Blunt 1958], and Blondel possessed the original version as well as a French translation made in the sixteenth century by Jean Martin, who also translated the works of Serlio, Vitruvius and Alberti [Gerbino 2002]. The architectural value of this 'strife of love in a dream' was immediately recognized by Colonna's contemporaries. In fact one can find there, years before Zorzi's speculations, not a simple comparison between architect and musician, but an assimilation of the method of design employed by the former to that used by the latter [Onians 1988: 207-215]:

> *Perche in alcuna parte havendo facto moto del fine debito allarchitectare, che e la præstante inventione, di acquistare modulatamente dil aedificio il solido corpo. Poscia licentemente quello invento, Lo Architecto perminute divisione el reduce, Ne piu ne meno quale il Musico havendo invento la intonatione & il mensurato tempo in una maxima quello da poi proportionando in minute Chromatice concinnamente sopra il solido lui el riporta. Per tale similitudine dapo la inventione la principale regula peculiare al Architecto e la quadratura. Et questa distribuentila in parvissime, La harmonia se gli offerisce dil aedificio & commodulatione, Et al suo principale gli convenienti correlarii* [Colonna 1499: fol. c iiii].

(This is why I have spoken in several places about the proper goal of architecture, which is its supreme invention: the harmonious establishment of the solid body of a building. After the architect has done this, he reduces it by minute divisions, just as the musician sets the scale and the largest unit of rhythm before subdividing them proportionately into chromaticisms and small notes. By analogy with this, the first rule that the architect must observe after the conception of the building is the square, which is subdivided to the smallest degree to give the building its harmony and consistency and to make the parts correlate with the whole [Colonna 1999: 47].)

Colonna's use of the word *concinnamente* shows that he was a careful reader of Alberti. (We do not need to resort to the excessive conclusions of Liane Lefaivre, who identified Colonna with Alberti despite all the philological, historical and lexicographical evidence to the contrary) [Schmidt 1978; Borsi 1995; Lefaivre 1997]. However, there is a fundamental difference in their respective approaches. Alberti had worked in an ontological and metaphorical model in which proportions (embodied for example in the Pythagorean *tetraktys*) reflect the harmony of cosmos; for him, this harmony should also stand as the basis for architecture. By contrast, Colonna's point of departure was a practical one, defined by the initial blueprint (the *preliminare quadratura*). The passage cited above is fundamental, for it reveals two important points: firstly, the evolution of architectural practice from the medieval model, which was based on the practical experience of the building site, to the renaissance model, which was based on a projectual design (the *cosa mentale*); and secondly, a conviction that the architect, when designing his project, must act in a way analogous to the composer. It is in Colonna's text that we first find this comparison made so clearly. The architect is no longer compared to the practical musician – the *cantor* of the medieval tradition – but rather to the learned theorist, the *musicus*, who manipulates both dimensions of *scientia musica* – the vertical axis (the *intonatione* of the intervals determined by the monochord) and the horizontal axis (the *mensurato tempo*, the relationships of prolation between the different temporal values) – and records what he had conceived mentally through notation. For Colonna, this is equivalent to the architect's preliminary *quadratura* [Zara 2010]. In my opinion, this change in approach reveals a switch to a linear concept of architecture, one which is not only anthropomorphic but more precisely rhetorical, linear and not volumetric, qualitative rather than quantitative.

In her study of the architectural representation contained in the *Hypnerotomachia Poliphili* and of the conceptual mechanisms that hold Colonna's descriptions, Roswitha Stewering points out the following:

Contrary to the anthropomorphistic conception of architecture, in which human proportions are considered the basis for the aesthetic norm, in the 'Hypnerotomachia' architectural theory seeks to incorporate human qualities: reason is manifested in harmonious proportions [Stewering 2000: 10].

This integration of human qualities into mathematical measure makes it possible to achieve an assimilation of the musical theory of affects and the anthropomorphic theory of the architectonic orders in a rhetorical key. Take for example the following passage: 'The place and situation seemed very difficult [*laborioso*]. Logistica accordingly, noticing this, immediately began to sing in the Dorian mode [*cum Dorio modo & tono di cantare*].' Here Colonna assigns a musical logic to the architectural one defined by Alberti for the Doric order, which he describes as 'rather suitable for work' (*ad laborem aptius*) [Colonna 1499: f. i]. A similar process happens to Venus, whose song is in the Lydian mode and whose the temple is built on Corinthian columns, the mode and order particular to the goddess of love [Onians 1988: 211-212]. This attitude reveals a change from a linear to a volumetric concept of harmony in architecture, a change from a quantitative mathematics based on *symmetria* and a qualitative mathematics based on *eurhythmia*. This shift is expressed by a textual amendment suggested in 1511 by Frà Giocondo to a contested passage near the beginning of the first book of Vitruvius' *De architectura*: *Ratiocinatio autem est quae res fabricatas sollertiae ac rationis pro portione demonstrare atque explicare potest*. Frà Giocondo amended *sollertiæ ac rationis pro portione* to *solertia ac ratione proportionis* [Vitruvius 1511: f. A1r-v]. This amendment – accepted by most sixteenth century editors, such as Georg Messerschmidt (1543), Guillaume Philandrier (1552) and Daniele Barbaro (1567) – gives the misleading impression that Vitruvius considered proportion as a guiding principle of his work right from the beginning of the first book, rather than being restricted to his discussion (in the third book) of the analogy between the proportions of architecture and those of the human body. While fifteenth-century commentators on Vitruvius seem to agree with the master's definition of architectural *ratiocinatio* – with construction (*fabrica*) as one of the two elements of architectural knowledge – Frà Giocondo's amendment gives quite a different impression of Vitruvius' project. This change of attitude is attested in the passage from Colonna, who clearly thinks of proportion as the basis of the *preliminare quadratura*. As Pierre Caye comments:

> Frà Giocondo, and Philandrier after him, does not read *proportione rationis*, but *proportionis ratione*, that is, 'according to the principle of proportion'. [If one reads the text with Giocondo's emendation,] the theory of proportions, that is the use of mathematics, that till then in Vitruvius occurred only in the third book (in the formal context of the description of the body and the description of figures), appears right from the first book as the primary operator of the architectural method of the conception of the project [Caye 2007 (my trans.)].

While Vitruvius works in terms of volume, the new focus on proportion, while already hinted at in Alberti, becomes fully evident in Colonna (thanks to the musical analogy), and comes to fuller expression in Giocondo's misguided emendation of Vitruvius, which would be widely accepted by subsequent sixteenth-century editors. Colonna does not use Alberti's terms *utilitas* and *pulchritudo*, but substitutes for them the terms *lineamento* and *prattica* respectively, which are taken not from Vitruvius' distinction beetwen *ratiocinatio* and *fabrica*, but are borrowed (in Italian translation) from the contemporary music theorist Franchino Gaffurio (*Theoricum opus musicae disciplinae* and *Practica musice*) [Onians 1988: 208]. For Colonna, the architect no longer uses the mechanical mediaeval process of design *ad quadratum* or *ad triangulum*; rather, by analogy with music, the architect obtains a law or a measure to which every part of the project is related (*lo Architecto perminute divisioni el reduce*). Palladio's letter to Giovan di Pepolli suggests that he may have been aware of the relation between the anthropomorphic paradigm and musical proportion under the sign of *eurhythmia*:

> *L'architettura non è altro che una proporzione dei membri in un corpo, cussì ben l'uno con gli altri e gli altri con l'uno simetriati e corrispondenti, che armonicamente rendino maestà e decoro [...] dee il corpo con membri*

e questi con quello aver insieme armonica proporzione, e che da quello nasce poi quel bello che da gli antichi greci Heurithmia [sic] vien detto: che altro non vuol dire che cussì ben composto corpo che più non si desideri [Portoghesi 2008: 177].

The final testimony to the rise of a rhetorical approach to music in the sixteenth century is the gradual application of an architectural feature – the staircase (*scala*) – to a basic musical phenomenon. Over the course of the sixteenth century, the term 'scale' gradually ousted the Greek term *systema*. The metaphor of the 'musical staircase' first appears in Andreas Ornithoparchus' *Musice Active Micrologus* (1517). In the second edition of his *Commento a Vitruvio* (1567), Barbaro felt it necessary to define this term. Zarlino manoeuvres it with caution, preferring the old term *sistema*. The presence of the word – and the metaphor – in the title of Ottavio Scaletta's treatise *Scala di musica necessaria per principianti* (1585) signals its definitive adoption [Giani 2000].

4 Conclusions

The new relationship between music and architecture, brought to maturity in the seventeenth century, illustrates a fundamental shift in the conceptualisation of architecture, a change that goes beyond analogical agreement and that touches directly on the foundation and the nature of architectural thought and its definition. This knowledge had been sketched in a speculative and literary way in the fifteenth century, was developed to some extent in a rhetorical conception of the harmonic parallels between music and architecture during the sixteenth century, and was at last completely encoded in the seventeenth. The narrative presented here thus poses a serious challenge to Wittkower's assertion that the analogy between architecture and music, laid down in the Italian Renaissance, broke down during the seventeenth century, degenerating particularly in France into a kind of formalism.

Many questions arise from this interpretation, which deserve more attention than can be offered here. First of all, in my opinion, is the spiritual legacy presented by the biblical description of Solomon's temple as an ideal archetype for Christian churches. The presence of this heritage is particularly evident in Villalpando and Ouvrard, in each case with some kind of musical connexion. The very universality of the Solomonic model led Perrault to ridicule his predecessors and offer a more auster and 'purified' version. Also deserving of more attention is the central figure of Sebastiano Serlio, who examined Zorzi's *memorandum* and later taught Delorme; Serlio is rightly described by Yves Pauwels from a rhetorical perspective as *Praeceptor Galliae* [Pauwels 2002]. Serlio's riddle-speculations on order and decoration, as well his hidden theological ideas, still divide historians [Tafuri 1985; Carpo 1993; Frommel 1998]; and as yet there has been no study of his musical knowledge. And we must take account, even in discussions of music, of the role of images in the transmission of theoretical knowledge [Collins Judd 2000]. A case in point is the possible relationship between Francesco di Giorgio Martino's design for the church of Santa Costanza, the design of the temple of Venus in Colonna's *Hypnerotomachia*, and Delorme's design for a church based perfectly on divine (or musical?) proportions. What is the relationship between these three designs? How did the design pass from one hand to another? Delorme perhaps knew the *Hypnerotomachia*, but had he seen Martino's design? What did these men know of their respective predecessors? The problem of the circulation of architectural images is one that is only beginning slowly to be studied in depth [Carpo 1998; 2001].

This historical revision is intended to pose a number of questions that might lead to a better understanding of the nature of the parallels between music and architecture. Such a reconsideration must take into account the fact that the analogical relationship that links the two disciplines consists not simply of the application of musical proportions to the spatial dimensions of a building, but also reflects fundamental epistemological shifts, such as the move from a consideration of abstract intervals to intervals between real sounds; or particular developments in the interpretation of canonical texts. Analogy is much more rich, nuanced, complex and fascinating.

Acknowledgments

First of all I wish to thank Marco Martin and Maria Semi, who from distant lands and with infinite kindness, translated the initial version of this essay. But, above all, my thanks and gratitude go to Grantley McDonald, who not only turned English from a second language into that of a native speaker, but also helped me, through questions and acute criticism, clarify several passages of the present work. If my line of argument is clear, it is thanks to him.

References

Allard, Joseph C. 1982. Mechanism, Music, and Painting in 17th Century France. *The Journal of Aesthetics and Art Criticism* **40**, 3: 269-279.

BARKER, Naomi Joy. 2000. 'Diverse Passions': Mode, Interval and Affect in Poussin's Paintings. *Music in Art* **25**, 1-2: 5-24.

BATTISTI, Eugenio. 1968. Le tendenze all'unità verso la metà del Cinquecento. *Bollettino del Centro Internazionale di Studi di Architettura Andrea Palladio* **10**: 127-146,.

———. 1973. Un tentativo di analisi strutturale del Palladio tramite le teorie musicali del Cinquecento e l'impiego di figure retoriche. *Bollettino del Centro Internazionale di Studi di Architettura Andrea Palladio* **15**: 211-232.

BERNIN DE SAINT-HILARION, François. 1680. *Des Proportions d'Architecture*. Münich, Bayerische Staatsbibliothek, Cod. Icon 193.

BLONDEL, François. 1675-1683. *Cours d'Architecture enseigné dans l'Académie Royale d'Architecture*. Paris: P. Abouin and F. Clouzier.

BLUNT, Anthony. 1937. The Hypnerotomachia Poliphili in the 17th Century France. *Journal of the Warburg Institute* **I**, 2: 117-127.

———. 1958. *Philibert De l'Orme*. London: Zwemmer.

BONHÔTE, Jean-Marc. 1971. Resonance musicale d'une ville de Palladio. *Musica Disciplina* **25**: 171-178.

BORSI, Stefano. 1995. *Polifilo architetto. Cultura architettonia e teoria artistica nell' 'Hypnerotomachia Poliphili' di Francesco Colonna, 1499*. Rome: Officina.

BRAULT, Yoann. 2006. La défense des proportions dans le 'Cours d'Architecture' de François Blondel. Pp. 30-38 in *Claude Nicolas Ledoux et le livre d'architecture – Étienne Louis Boullée l'utopie et la poésie de l'art*, Daniel Rabreau and Dominique Massounie eds. Paris: Monum.

BRENET, Michel [Bobillier, Marie]. 1910. *Les Musiciens de la Sainte-Chapelle du Palais*. Paris: Picard.

BRISEUX, Charles-Estienne. 1752. *Traité du beau essentiel dans les arts, Appliqué particulierement à l'Architecture, et demontré Phisiquement et par l'Expérience; Avec Un traité des Proportions Harmoniques, et l'on fait voir que c'est de ces seules Proportions que les Édifices généralement approuvés, empruntent leur Beauté réelle et invariable*. Paris: Chez l'Auteur et Chereau.

BUCH, David J. 1989. The Coordination of Text, Illustration, and Music in a Seventeenth-Century Lute Manuscript: 'La Rhétorique des Dieux'. *Imago Musicae* **6**: 39-81.

CAMPANINI, Saverio. 2007. Francesco Zorzi: armonia del mondo e filosofia simbolica. Pp. 239-260 in *Il pensiero simbolico nella prima età moderna*, Annarita Angelini and Pierre Caye, eds. Florence: Olschki.

CAYE, Pierre and CHOAY Françoise. 2004. *Leon Battista Alberti: L'Art d'Édifier*. Paris: Seuil.

CAYE, Pierre. 2007. L'édition du 'De architectura' de Vitruve et la constitution du savoir architecturale à la Renaissance, comunicazione letta al Colloquio Internazionale L' 'Archivium' et le travaille de la pensée. Humanisme philologique, humanisme philosophique, Paris, 22-23 May 2007.

CARPO, Mario. 1993. *La maschera e il modello. Teoria architettonica ed evangelismo nell' 'Extraordinario Libro' di Sebastiano Serlio.* Milan: Jaca Book.

———. 1998. *L'architettura dell'età della stampa. Oralità, scrittura, libro stampato e riproduzione meccanica dell'immagine nella storia delle teorie architettoniche.* Milano: Jaca Book (Eng. trans. *Architecture in the Age of Printing. Orality, Writing, Typography, and Printed Images in the History of Architectural Theory*, Sarah Benson, trans. Cambridge, MA: The MIT Press, 2001).

———. 2001. How Do You Imitate a Building That You Have Never Seen? Printed Images, Ancient Models, and Handmade Drawings in the Renaissance Architectural Theory. *Zeitschrift für Kunstgeschichte* **64**, 2: 223-233.

COHEN, Albert. 1974. René Ouvrard (1624-1694) and the beginnings of French Baroque Theory. *Report of the Eleventh ISM Congress – Copenhagen 1972.* Copenhagen: Wilhelm Hansen, vol. I, pp. 336-342.

———. 1975. The Ouvrard-Nicaise correspondence (1663-93). *Music & Letters* **56**, 3-4: 356-363.

COLONNA, Francesco. 1499. *Hypnerotomachia Poliphili.* Venice: In aedibus Aldi Manutii.

———. 1999. *Hypnerotomachia Poliphili. The Strife of Love in a Dream.* Joscelyn Godwin, trans. London: Thames & Hudson.

COLLINS JUDD, Cristle. 2000. *Reading Renaissance Music Theory. Hearing with the Eyes.* Cambridge: Cambridge University Press.

COSTANTINI, Michela. 2002. La trasformazione storica dell'applicazione dei rapporti musicali all'architettura attraverso la lettura armonica della base attica. *Le culture della tecnica*, nuova serie **14**: 75-102.

DELORME, Philibert [Philibert de l'Orme]. 1567. *Premier tome de l'architecture de Philibert De l'Orme conseiller et aumônier ordinier du Roi, et abbé de S. Serge les Angiers.* Paris: Federic Morel.

DIONISOTTI, Carlo. 1968. *Gli umanisti e il volgare fra Quattro e Cinquecento.* Florence: Le Monnier.

FICHET, Françoise. 1979. *La théorie architecturale à l'âge classique. Essai d'anthologie critique.* Liège: Mardaga.

FOSCARI, Antonio and TAFURI, Manfredo. 1983. *L'armonia e i conflitti: la chiesa di San Francesco della Vigna nelle Venezia del 500.* Torino: Einaudi.

FRÉART DE CHAMBRAY, Roland. 1650. *Parallèle de l'Architecture Antique et Moderne, avec un Recueil des Six Principaux Autheurs qui ont écrit des Cinq Ordres, savoir: Palladio et Scamozzi, Serlio et Vignola, D. Barbaro et Cataneo, L. B. Alberti et Viola, Bullant et De L'Orme, comparés entre eux. Par R. Fréart, sieur de Chambray.* Paris: E. Martin. (Modern ed.: *Parallèle de l'architecture antique avec la moderne suivi de Idée de la perfection de la peinture*, Frédérique Lemerle-Pauwels and Milovan Stanic, eds. Paris: Ecole Nationale Supérieure des Beaux-Arts, 2005.)

FROMMEL, Sabine. 1998. *Sebastiano Serlio architetto.* Milan: Electa.

GERBINO, Anthony. 2002. The Library of François Blondel 1618-1686. *Architectural History* **45**: 289-324.

———. 2010. *François Blondel: Architecture, Erudition, and the Scientific Revolution.* London & New York: Routledge.

GIANI, Maurizio. 2000. Scala musica. Vicende di una metafora. Pp. 31-48 in *Le parole della musica. III. Studi di lessicologia musicale*, Fiamma Nicolodi and Paolo Trovato, eds. Florence: Olschki.

GIOSEFFI, Decio. 1972. Il disegno come fase progettuale dell'attività palladiana. *Bollettino del Centro Internazionale di Studi di Architettura Andrea Palladio* **14**: 45-62.

———. 1978. Dal progetto al trattato: incontro e scontro con la realtà. *Bollettino del Centro Internazionale di Studi di Architettura Andrea Palladio* **20**: 27-45.

———. 1980a. I disegni dei 'Quattro Libri' come modelli: modellistica architettonica e teoria dei modelli. *Bollettino del Centro Internazionale di Studi di Architettura Andrea Palladio* **22**: 147-64.

———. 1980b. Convegno palladiano: precisazioni dovute. *Bollettino del Centro Internazionale di Studi di Architettura Andrea Palladio* **22**, 2: 193-203.

———. 1989. Palladio oggi: dal Wittkower al postmoderno. *Annali di architettura* **1**: 105-121.

GODMAN, Rob. 2008. The Enigma of Vitruvian Resonating Vases and the Relevance of the Concept for Today. *International Computer Music Conference Proceedings*, Queen's University, Belfast. Available at: http://quod.lib.umich.edu/cgi/t/text/pagevieweridx?c=icmc;idno=bbp2372.2008.065;cc=icmc; view=image (last accessed 3 January 2011).

GOZZA, Paolo. 2007 (2009). I suoni taumaturghi. Un'estetica musicale barocca dello spossessamento. *Musica e Storia* **15**, 2: 417-441.

HERRMANN, Wolfgang. 1967. Unknown designs for the 'Temple of Jerusalem' by Claude Perrault. Pp. 143-158 in *Essays in the History of Architecture presented to Rudolf Wittkower*, Douglas Fraser, Howard Hibbard and Milton J. Lewine, eds. London: Phaidon.

———. 1973. *The theory of Claude Perrault*. London: Zwemmer.

HERSEY, George L. 1976. *Pythagorean Palaces. Magic and Architecture in the Italian Renaissance*. Ithaca: Cornell University Press.

———. 2000. *Architecture and Geometry in the Age of the Baroque*. Chicago: University of Chicago Press.

HOWARD, Deborah and Malcolm LONGAIR. 1982. Harmonic Proportion and Palladio's 'Quattro Libri'. *Journal of the Society of Architectural Historians* **41**, 2: 116-143

JOUANNY, Charles. 1911. *Correspondance de Nicolas Poussin*. Paris: Jean Schemit.

KAPPRAFF, Jay and Ernest G. MCCLAIN. 2005. The System of Proportions of the Parthenon: A Work of Musically Inspired Architecture. *Music in Art* **30**, 1-2: 5-16.

LE CAMUS DE MÉZIÈRES, Nicolas. 1780. *Le génie de l'architecture; ou, L'analogie de cet art avec nos sensations*. Paris: B. Morin. (Eng. trans. *The Genius of Architecture; or, The Analogy of That Art with Our Sensations*, Robin Middleton, ed. Chicago: University of Chicago Press, 1992.)

LEFAIVRE, Liane. 1997. *Leon Battista Alberti's Hypnerotomachia Poliphili. Re-Cognizing the Architectural Body in the Early Italian Renaissance*. Cambridge, MA: MIT Press.

LEMERLE, Frédérique. 2006. Claude Perrault théoricien: l' 'Ordonnance des Cinq Espèces de Colonnes' (1683). Pp. 18-29 in *Claude Nicolas Ledoux et le livre d'architecture – Étienne Louis Boullée l'utopie et la poésie de l'art*, Daniel Rabreau and Dominique Massounie eds. Paris: Monum.

LEMONNIER, Henry. 1912. *Procès-Verbaux de l'Académie Royale d'Architecture 1671-1793*. 10 vols (1911-1929). Paris, Jean Schemit.

LOCKSPEISER, Edward. 1967. Poussin et les modes. *Revue de Musicologie* **53**: 61-64.

MAMBELLA, Guido. 2008. Corpo sonoro, geometria e temperamenti. Zarlino e la crisi del fondamento numerico della musica. Pp. 185-233 in *Music and Mathematics in Late Medieval and Early Modern Europe*, Philippe Vendrix ed. Turnhout: Brepol.

MARCH, Lionel. 1998. *Architectonics of Humanism: Essays on Number in Architecture*. London: Wiley-Academy.

———. 2008. Palladio, Pythagoreanism and Renaissance Mathematics. *Nexus Network Journal* **10**, 2: 227-244.

MARTIN, Marie-Pauline. 2006. L'analogie des proportions architecturales et musicales: évolution d'une stratégie. Pp. 40-47 in *Claude Nicolas Ledoux et le livre d'architecture – Étienne Louis Boullée l'utopie et la poésie de l'art*, Daniel Rabreau and Dominique Massounie eds. Paris: Monum.

MÉROT, Alain. 1994. Les modes, ou les paradoxes du peintre. Pp. 80-86 in *Nicolas Poussin 1594-1665*, exhibition catalogue, Pierre Rosenberg and Louis-Antoine Prat eds. Paris: RMN.

MILLON, Henry A. 1972. Rudolf Wittkower, 'Architectural Principles in the Age of Humanism': Its Influence on the Development and Interpretation of Modern Architecture. *Journal of the Society of Architectural Historians* **31**, 2. 83-91

MITROVIC, Branko. 1990. Palladio's Theory of Proportions and the Second Book of the 'Quattro Libri dell'Architettura'. *Journal of the Society of Architectural Historians* **49**, 3: 279-292.

———. 1998. Paduan Aristotelianism and Daniele Barbaro's Commentary on Vitruvius 'De Architectura'. *Sixteenth Century Journal* **29**, 3: 667-688.

———. 1999. Palladio's Theory of the Classical Order and the First Book of 'I Quattro Libri dell'Architettura'. *Architectural History* **42**, 2: 110-140.

———. 2001. A Palladian Palinode: Reassessing Wittkower's 'Architectural Principles in the Age of Humanism'. *Architectura* **31**, 2: 113-131.

———. 2004. *Learning from Palladio*. New York: Norton.

MOURJOPOULOS, Vassilantopopoulos. 2003. A study of Ancient Greek and Roman Theater. *Acta Acoustica* **89**: 123-136.

NAREDI-RAINER, Paul von. 1977. Musikalische Proportionen, Zahlenästhetik und Zahlensymbolik im architektonischen Werk L. B. Albertis. *Jahrbuch des Kunsthistorischen Institutes der Universität Graz* **12**: 81-213.

———. 1982. *Architecktur und Harmonie. Zahl, Maß und Proportion in der abendländischen Baukunst*. Köln: DuMont.

———. 1985. Musiktheorie und Architektur. Pp. 149-174 in *Geschichte der Musiktheorie. I. Ideen zu einer Geschichte der Musiktheorie*, Frieder Zaminer ed. Darmstadt: Wissenschaftliche Buchgesellschaft.

———. 1994. La bellezza numerabile: l'estetica architettonica di Leon Battista Alberti. Pp. 292-299 in *Leon Battista Alberti*, Joseph Rykwert and Anne Engel eds. Milan: Olivetti-Electa.

ONIANS, John. 1988. *Bearers of Meaning. The Classical Orders in Antiquity, the Middle Ages and the Renaissance*. Princeton: Princeton University Press.

OUVRARD, René. 1679. *Architecture Harmonique, ou l'Application de la Doctrine des Proportions de la Musique à l'Architecture*. Paris: J. B. de la Caille.

PAYNE, Alina A. 1994. Rudolf Wittkower and Architectural Principles in the Age of Modernism. *Journal of the Society of Architectural Historians* **53**: 322-342.

PAUWELS, Yves. 2000. 'Harmonia est discordia concors': le modèle musical dans l'architecture des temps modernes. Pp. 313-325 in *L'Harmonie*, Charles Charraud ed. Orléans: Meaux.

———. 2002. *L'architecture au temps de la Pléiade*. Paris: Monfort.

PÉROUSE DE MONTCLOS, Jean-Marie. 2000. *Philibert de l'Orme. Architecte du Roi (1514-1570)*. Paris: Mengè.

PICON, Antoine. 1989. *Claude Perrault ou la curiosité d'un classique*. Paris: Picard.

PINTORE, Angela. 2004. Musical Symbolism in the Works of Leon Battista Alberti. From 'De re aedificatoria' to the Ruccelai Sepulchre. *Nexus Network Journal* **6**, 2: 49-70.

POMME DE MIRIMONDE, Albert. 1972. Poussin et la musique. *Gazette des Beaux-Arts* **79**: 129-150.

PORTOGHESI, Paolo. 2008. *La mano di Palladio*. Torino: Allemandi.

RAMÍREZ, Juan Antonio, ed. 1991. *Dios Arquitecto. J. B. Villalpando y el Templo de Salómon*. Madrid: Siruela.

RYKWERT, Joseph. 1996. *The Dancing Column: On Order of Architecture*. Cambridge, MA: MIT Press.

ROBISON, Elwin C. 1998-99. Structural Implications in Palladio's Use of Harmonic Proportions. *Annali di architettura* **10-11**: 175-182.

SAMSA, Danilo. 2003. L'Alberti di Wittkower. *Albertiana* **6**: 51-94.

SANCHEZ DE ENCISO, Sabina. 2008. Música y arquitectura en el 'De postrema Ezechielis prophetae visione' de J.B. Villalpando. *Cuadernos de Música Iberoamericana* **15**: 7-40.

SERRAVEZZA, Antonio and Paolo GOZZA. 2004. *Estetica 'e' musica. L'origine di un incontro*. Bologna: CLUEB.

SCALVINI, Maria Luisa and Sergio VILLARI. 1991. *Claude Perrault: L'ordine dell'architettura*. Palermo: Aesthetica Preprint.

———. 1994. *Il manoscritto sulle proporzioni di François Bernin de Saint-Hilarion*. Palermo: Aesthetica Preprint.

SCHMIDT, Dorothea. 1978. *Untersuchungen zu den Architekturekphrasen in der Hypnerotomachia Poliphili. Die Beschreibung des Venus-Tempels.* Frankfurt am Main: R. G. Fischer.

STEWERING, Roswitha. 2000. Architectural Representations in the 'Hypnerotomachia Poliphili' (Aldus Manutius, 1499). *Journal of the Society of Architectural Historians* **59**: 16-25.

STUMPFHAUS, Bernhard. 2005. *Modus – Affekt – Allegorie bei Nicolas Poussin.* Reimer: Dietrich.

SZAMBIEN, Werner. 1986. *Symétrie Goût Caractère. Théorie et terminologie de l'architecture à l'âge classique 1550-1800.* Paris: Picard.

TAFURI, Manfredo. 1985. *Venezia e il Rinascimento. Religione, scienza, architettura.* Torino: Einaudi.

TESTA, Fausto. 2003. Il 'Traité du Beau Essentiel' di C.-E. Briseux e il tema delle proporzioni armoniche nella teoria architettonica del secolo dei Lumi. *Oltrecorrente* **7**, 1: 143-154.

VASOLI, Cesare. 1998. Il tema musicale e architettonico della 'Harmonia Mundi' da Francesco Giorgio veneto all'Accademia degli Uranici e a Gioseffo Zarlino. *Musica e Storia* **6**, 1: 193-210.

VENDRIX, Philippe. 1988. René Ouvrard et l'évolution de l'art musical. *Revue belge de musicologie* **42**: 193-197.

———. 1989. Proportions harmoniques et proportions architecturales dans la théorie française des XVIIᵉ et XVIIIᵉ siècles. *International Review of the Aesthetics and Sociology of Music* **20**, 1: 3-10.

———. 1992. L'augustinisme musical en France au XVIIᵉ siècle. *Revue de Musicologie* **78**: 237-255.

VERGO, Peter. 2005. *That Divine Order. Music and the Visual Arts from Antiquity to the Eighteenth Century.* London: Phaidon.

VILLALPANDO, Juan Bautista and Jeronimo DEL PRADO. 1595-1604. *Ezechielem Explanationes et Apparatus Urbis ac Templi Hierosolymitani*, Rome: A. Zanetti.

VITRUVIUS. 1511. *M. Vitruvius per Iocundum solito castigatior factus cum figuris et tabula ut iam legi et intelligi possit.* Venice: Giovanni Tacuino de Tridino.

———. 1543. *M. Vitruvii viri svæ professionis peritissimi, de architectura libri decem, ad Augustum Cæsarem accuratiss. conscripti: & nunc primum in Germania qua potuit diligentia excusi, atque hinc inde schematibus non iniucundis exornati.* Strasbourg: Knobloch-Machæropiœus [Messerschmidt].

———. 1552. *M. Vitruvii Pollionis de architectura libri decem ad Cæsarem Augustum, omnibus omnium editionibus longè emendatiores, collatis veteribus exemplis*, Guillaume Philandrier, ed. Lyon: Jean de Tournes.

———. 1567. *M. Vitruvii Pollionis de architectura libri decem, cum commentariis Danielis Barbari, electi Patriarchæ Aquileiensis.* Venice: Franciscus Franciscium Sensensis & Ioannes Crugher Germanus.

WITTKOWER, Rudolf. 1949. *Architectural Principles in the Age of Humanism.* London: Warburg Institute. (Citations in this present paper taken from the 3rd ed., London: Academy Editions, 1962.)

ZANONCELLI, Luisa. 1999. Reciproche influenze dell'idea di 'divina proporzione'. Pp. 199-212 in *Leon Battista Alberti. Architettura e cultura*, Claudio Gallico ed. Florence: Olschki.

———. 2007. La musica e le sue fonti nel pensiero di Leon Battista Alberti. Pp. 85-116 in *Leon Battista Alberti teorico delle arti e gli impegni civili del 'De re aedificatoria'*, Arturo Calzona, Francesco Paolo Fiore, Alberto Tenenti and Cesare Vasoli eds., 2 vols. Florence: Olschki.

ZARA, Vasco. 2005. Musica e Architettura tra Medio Evo e Età moderna. Storia critica di un'idea. *Acta Musicologica* **77**, 1: 1-26

———. 2006. Antichi e Moderni tra Musica e Architettura. All'origine della 'Querelle des Anciens et des Modernes'. *Intersezioni* **26**, 2: 191-210.

———. 2007. Da Palladio a Wittkower. Questioni di metodo, di indagine e di disciplina nello studio dei rapporti tra musica e architettura. Pp. 153-190 in *Prospettive di iconografia musicale*, Nicoletta Guidobaldi ed. Milan: Mimesis.

———. 2008a. Una storia della musica all'ombra di Port-Royal: la 'Musique Rétablie' di René Ouvrard. *Studi Musicali* **37**, 1: 59-100

————. 2008b. Modes musicaux et ordres d'architecture: migration d'un modèle sémantique dans l'œuvre de Nicolas Poussin. *Musique – Images – Instruments. Revue française d'organologie et d'iconographie musicale* **10**: 62-79

————. 2007 (2009). Suono e carattere della base attica. Itinerari semantici d'un metafora musicale nel linguaggio architettonico francese del Settecento. *Musica e Storia* **15**, 2: 443-474

————. 2010. Dall' 'Hypnerotomachia Poliphili' al Tempio di Salomone. Modelli architettonico-musicali nell' 'Architecture Harmonique' di René Ouvrard, 1679. Pp. 131-156 in *La réception de modèles 'cinquecenteschi' dans la théorie et les arts français du XVIF siècle*, Sabine Frommel and Flaminia Bardati, eds. Geneva: Droz.

ZARLINO, Gioseffo. 1558. *Le istitutioni harmoniche.* Venetia: Francesco de' Franceschi.

About the author

Vasco Zara is Maître de conférences in ancient music at the Université de Bourgogne of Dijon, researcher at the Unité Mixte de Recherche 5594 « ARTeHIS » of the Université de Bourgogne, and associated member of the Centre d'Études Supérieures de la Renaissance (Tours). He obtained his Ph.D. from the University of Bologna and from the Centre d'Études Supérieures de la Renaissance (Tours), with a dissertation on the works and theory of René Ouvrard, maître de musique at the Sainte-Chapelle in Paris during the second half of the Sixteenth-Century (and whose correspondence and *Architecture Harmonique, ou l'Application de la Doctrine des Proportions de la Musique à l'Architecture*, will be published shortly). He is one of the directors, together with Philippe Vendrix (CESR, Tours) and Ennio Stipcevic of Zagreb's Academy of Sciences, of the research program Renaissance Music in Croatia (http://ricercar.cesr.univ-tours.fr/3-programmes/EMN/Croatie). An affirmed scholar in the studies that deal with the relationship between music and architecture during the Medieval and the Renaissance periods, he is the bibliographical referee for the Study Group on Musical Iconography about these topics. His actual research focuses on the relations between polyphony, geometry and theology in the fourteenth and fifteenth centuries.

Lone Mogensen

Blidvädersvägen 6 G
S-222 28 Lund
SWEDEN
mogensenlone@hotmail.com

Keywords: Church, basilica, narthex, atrium, royal mansion, Scania, Denmark, Sweden, Medieval architecture, Euclidean geometry, ad quadratum, modules, measuring systems, descriptive geometry, symbolism

Research

Making a Difference

Abstract. In attempting to identify how the ground plans of medieval churches were initially staked out, it has been noted that the width of a given element (an aisle or a pier) equals the difference between two given distances in a square appearing in that same church; in a church plan constructed on a square module with the side 1 and the diagonal $\sqrt{2}$, several distances in that plan would be produced, such as $\sqrt{2}-1$ and $1-\sqrt{2}/2$, a fact that is rather well known. However, the difference between those two measurements has been used. How did the original constructor actually create that difference of differences onsite? The answer was found thanks to a plan of the Dalby church (ca. 1060) and a little cord which simply needed to be folded.

The folded cord

This article explains how the original constructor – the medieval master builder – once could have staked out the plan of a church on the ground. Of course it remains hypothetical even if it is sometimes expressed as a fact to give the explanation below a better flow.

In attempting to identify how the plans of medieval churches were initially staked out, I have found that the width, for instance of an aisle or a pier, equals the difference between two given distances in a square appearing in that same church. Take for instance a church plan constructed on a square module with the side 1 and the diagonal $\sqrt{2}$: several distances in that plan would be produced, such as $\sqrt{2}-1$ and $1-\sqrt{2}/2$. This is rather well known. But I also discovered that the difference between those two measurements has been used. That special distance I call x here .

How did the original constructor actually create that difference of differences onsite?

Simply by examining a church plan I was not able to find an easy method to produce the distance x. In order to improve my chances at understanding how the constructor had created this distance I used a plan of the Dalby church, which dates from around 1060, to scale 1:50. When I used a little cord made to the same scale as the church plan and tried to stake out the plan, the answer immediately revealed itself: the cord needed simply to be folded.

History

Today Dalby is a little village located in Scania in what is now the southernmost region in Sweden, but during the middle ages Scania was a part of Denmark. Around 1060, the Danish King Sven Estridsen was, as one medieval text put it, "the first to build Dalby church".[1] There is a general consensus among the Dalby researchers that his church is the one in question. Some generations later, the Anglo-Saxon Benedictine monk Aelnoth mentions the Dalby site again, though not as a church but as "the famous place".[2]

King Sven continued his predecessors' missionairy work in Denmark, and he installed Bishop Egino in Dalby. Egino probably had no bishopry of his own, since the dioscese of Scania belonged to the Cathedral ten kilometers away in Lund. The German Archbishop Adam of Bremen described him as a learned man and a successful missionairy bishop in Sweden and Denmark [Adam of Bremen 2000: 205]. Egino could have been the King's chancellor as well. In 1066 the Bishop in Lund died and Egino took his place. Upon his departure from Dalby he founded a college of canons which shortly after established an Augustinian monastery. Nevertheless, the Danish king kept his formal patrimony over the place, which grew to be very prestigious, independent, and wealthy.

King Sven's original basilica at Dalby was soon elongated eastwards, and a westwork, with two towers flanking a square middle part, built, although the towers were probably never completed. A lot of building activity took place on the site during the medieval ages, in part due to the growing prosperity of the monastery, but also as a result of devastating fires and war.

After the Reformation the site was impoverished by the King's favorite men whom he installed there to govern the famous place. The decay escalated when the region became Swedish in the seventeenth century. Today the choir, both the original as well as the elongated, and the eastern bay of the nave are gone, and so is the complete northern aisle. Of the twelfth century's westwork the two flanking towers are long gone. Only the bottom and part of the second floor remain of its middle part. It carries today the remnants of a later tower (fig. 1).

Fig. 1. The Dalby church. Photo by Knut Andreassen

During the twentieth century several archaeological excavations took place both inside and outside the church. The ground plan of the original basilica was documented, and the southwest corner of an older and original western part was found; it had most likely been a narthex.[3] The really big surprise was found in the area west of the church: just under the ground surface the foundations and lower courses of two houses were discovered. These have, for good reason, been interpreted as a royal mansion consisting of a main house flanked by a southern wing.

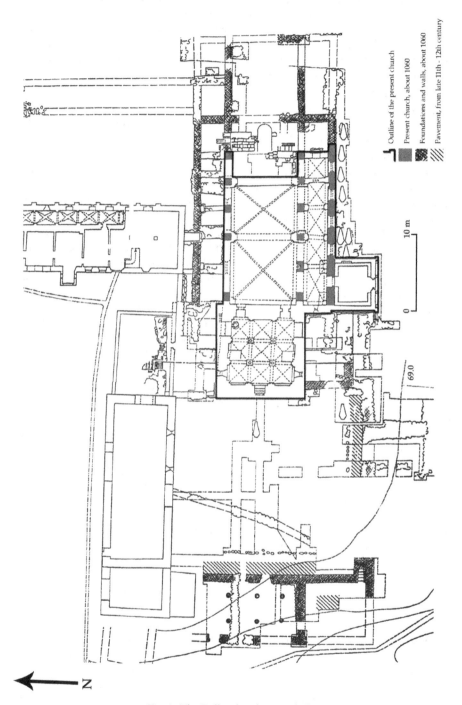

Fig. 2. The Dalby church ground plan

Outline of the present church
Present church, about 1060
Foundations and walls, about 1060
Pavement, from late 11th – 12th century

10 m

0

69.0

N

The main house, which had a cellar, lay parallel to the narthex, and was more or less connected to it by the southern wing. A probable northern wing could not be identified, because that place is occupied by a complicated house with medieval origin. If a northern wing had originally been built there, the three two-storey houses would have, together with the narthex, formed an atrium (fig. 2). Nothing like this has been found in Scandinavia so far. No wonder it was called "the famous place."

Staking out the plan

On a special day in the middle of the eleventh century the orientation of the basilica would have been laid out. The location as well as the planned construction work would have been blessed in a ceremony. When the roads and bridges to the quarry nearby were reinforced, the well and the water pipe functioning, timber, iron, lime and food could be supplied; when the lodge, the craftsmen and workers were ready, the site cleared and levelled, the very erection of the walls could begin.

Prior to that, however, the plan would first need to become visible on the ground. On the site itself or in his home the master builder (whoever he was) has constructed a square module. He would have used thin ropes, perhaps made of waxed hemp. The side of the square equals 1 (in Dalby this is equal to 25,6 meters; more about this below). The diagonals are $\sqrt{2}$. He then selects one of the diagonals as a ruler, this will be – itself or the model for – a special cord that resists extension caused by either its own weight, or by water and wind. On this diagonal cord he marks out the three distances $1/2$, $\sqrt{2}/2$ and 1. He could have used knots as marks, but to be more precise it would be better to sew threads of different colors or metals to the rope. Here, I shall refer to these marks as "knots".

When he marks the distance 1 on the diagonal cord, this mark will divide the second half of the diagonal in two pieces: $y = (\sqrt{2} - 1)$ and $z = (1 - \sqrt{2}/2)$. Then, using the cord alone out of the geometrical context, i.e., outside this square, when he folds it at point B (the distance 1 on the square between y and z), the end of the rope will at once mark the difference between y and z, which is the distance x. In other words: $x = (\sqrt{2} - 1) - (1 - \sqrt{2}/2)$. Of course all these distances can be found by a series of inscribed squares, but this is a handier shortcut (see fig. 3).

In Dalby the cord would probably have been at least $\sqrt{3}$ long, but here in these figures I take it to be $\sqrt{2}$, for simplification. This cord can also construct right angles. When using the distance $(x + z)$ as a base line fixed with pegs, and turning the cord ends up to meet, an isoceles triangle with a right angle at B is made.[4] Along with the cord, the master builder would also have had a couple of rulers. It would be most convenient to use a ruler of 2.56 meters and its fractions, along with a ruler with the applied yard and foot, and their divisons.

By use of the distances between the knots of this cord, a plan designed in this way would turn out to be identical with the Dalby buildings. The module 1 is 25.6 meters, $\sqrt{2}$ equals 36.2 meters, and $\sqrt{3} = 44.34$ meters. The distance x $(= 3/\sqrt{2} - 2)$ is therefore 3.1 meters. This distance is the exact width of the still existing south aisle, measured near the floor, close to the original ground surface where the staking out threads would have run. The nave together with its two framing arcade walls are 10.6 meters, which equals the distance y $(= \sqrt{2} - 1)$. Discrepancies are generally ± 0.5 cm or less. The excavation survey made to scale 1:50 shows a perfect match: the outlines of the walls follow the lines of this reconstruction. There is one irregularity, though, to which I shall return later. As the figures show, the staking out also includes the two buildings west of the church.

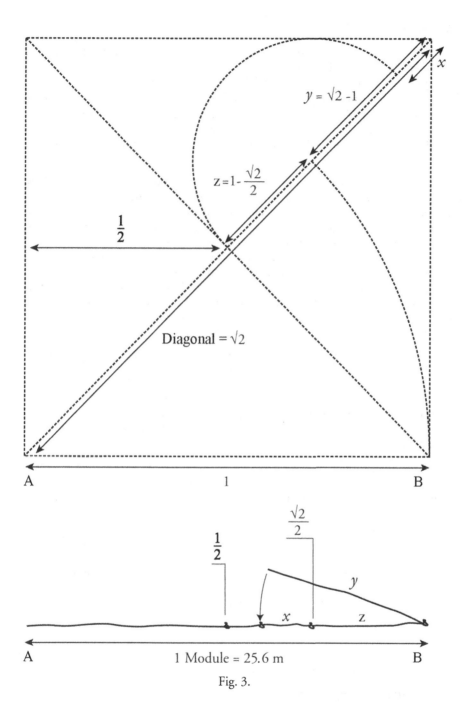

$y = \sqrt{2} - 1$

$z = 1 - \dfrac{\sqrt{2}}{2}$

$\dfrac{1}{2}$

x

Diagonal $= \sqrt{2}$

A   1   B

$\dfrac{1}{2}$

$\dfrac{\sqrt{2}}{2}$

y

x z

A   1 Module = 25.6 m   B

Fig. 3.

The module

Fifteen years ago I studied the plan of Lund Cathedral as well as the building itself [Mogensen 2003]. I found a module of 25.6 meters, even if some corners of the northern nave and transept are oblique. That module is the width of the nave, between the aisle's doors, and the longitudinal distance in the crypt. The width of the crypt is $\sqrt{2}$ x 25.6 meters and the length of the nave is $\sqrt{3}$ x 25.6 meters. The differences between differences are found in the width of pillars, walls and wall segments, which all can be derived from the square module shown here and a series of inscribed squares, or simply a folded cord.

In the medieval plan of Lund I find a structure with a module of 256 meters.[5] This is based on a modern map made to scale 1:1000, combined with maps of the reconstructed medieval street system, which is generally still in existence.[6] Also, a very early stone church in Lund, Drotten, seems to be related to the distance 25.6 meters. The first streets, blocks and churches were laid out in the period between 990 – 1020 when the Kings Harold Bluetooth, Sven Forkbeard, Hardecnut and Cnut the Great ruled. Chronologically, the city plan is oldest, followed by Drotten, the Dalby setting and then the present Lund cathedral, which is mainly from the beginning of the twelfth century. So far, I haven't found this module of 25.6 meters in other places, except in a publication about the church Østermarie on the island of Bornholm close to Scania, written by Erling Haagensen [1993: 159].

Haagensen has highlighted that this distance corresponds with several ancient measuring systems, among them an English mile: a circle with a radius of 256 meters will have a perimeter of an English mile. In fact, the radius must be a little longer, i.e., 256.032 meters, to give an exact mile. However, when I started my study I knew nothing about any given distance at all, and I did not recognize less than half a centimeter, so I stuck with 25.6 meters. Now knowing about the extra 0.32 centimeters, I still find them without importance in *this* context.[7]

It is tempting to look at plans published in books and on the Internet in order to discover this module in other buildings, ancient as well as medieval. However, in order to be sure, one has to take measurements onsite to be able to compare them with reliable drawings made to a large scale. I would therefore be very thankful if anyone else has identified this module in other contexts. The Danish architecture professor Mogens Koch has shown how the cubit, or rather the *ell*, used in the Lund Cathedral is the "natural cubit" of 46.666… centimeters, with a foot of 31.111… centimeters.[8] Most likely the same ell was used in Dalby, because the total length of the church was either 99 or 100 such ells, depending on whether or not there was a forepart on the narthex. At least in Lund and Dalby the natural cubit seems to be combined with the module 25.6 meters. Despite these commonalities no general measurement has been identified with certainty in buildings from that period in Scandinavia.

However, whatever the module was, I did find evidence that points to the existence of a folded cord with the differences mentioned above.

Mistakes and humble gestures

There is sometimes a tendency to explain irregularities in church construction as the result of humble gestures by building masters. According to this explanation, the master mason deliberately made an "error" to demonstrate that nobody except the architect of the cosmos, God himself, is able to create a perfect construction. Even if the church symbolized the cosmos and the act of building a church was a humble imitation of God's great act of creation (hence the profound symbolism in the building process), there was however a blasphemous contradiction in such a gesture: if the mason did *not* choose to make his work imperfect, he would in fact equal the Creator in designing perfection. Furthermore, to construct a defect and integrate all its consequences in an otherwise perfect system – that requires even more skill. Maybe there was a *gestus humilitatis* that was sophisticated in its mere simplicity?

Most plans suffer from some imperfection which could have forced the masons to adjust both small details and the orientation of big walls. Here I present two sorts of failures which can be traced to a specific step in the setting out process, a step that the master could choose to take, or that he might easily do as a genuine mistake.

In the Dalby plan, a perfect way to stake out the width of the total nave is to use the distance √2/2, as the second step of fig. 4. That has apparently happened in the northeast corner. Here the total bredth of the building from point B to C appears to be 18.10 meters which equals √2/2 distance. But using that method to determine the north wall in its whole length would completely destroy the symmetry of the whole. So, to make it work, the northern wall now deviates from that corner (B) and points inwards as it extends towards west, narrowing off the distance at the western side by at least 35 centimeters.[9] This deviation mainly affects the width of the northern aisle and its northern wall, and hardly anything on the nave arcade.

If the mason deliberately made this irregularity, it might express something like this: "Look, how gracefully it could have been done! But alas, only the migthy Creator is able to master such an elegant simplicity in His methods. I have to do it in a more longwinded way, by adding distance to distance….".

Another sort of irregularity is the switching side of a wall. It can be seen in the plan of Lund Cathedral as well as in many other churches, among them the southern transept wall in St. Etienne in Nevers, which has previously been described in this journal [Zenner 2002].[10] Imagine the latter building site filled with stakes, between which a certain amount of strings are drawn to show the workers where to dig the foundation pits. Even if only one of the two outlines of the walls were drawn at once, the building site would have looked like a spider's web at such a stage. If, for instance, the inside of the western wall of the northern transept was important in the staking out process, a string must have been extended there, and might have continued over the middle crossing to show the corresponding outline of the southern transept. When the width of this wall was to be depicted with a parallel thread, how easy would it be to pick the wrong side of the first string?

The consequences of such a mistake, deliberate or not, can be seen as an oblique angle in a couple of corners, and a mysterious deficiency or excess of distance, emerging everywhere.

I believe that after having staked out the main structure of a church plan, the remainder was laid with either a measuring rod, such as the stick of 2.56 meters I guess was used in Dalby, which also is the space between the eastern arcade pillars, or without any tools at all, and only done by the eye. The Dalby basilica had a simple design, and the walls in the present church are still extremely straight so both sides of the walls might very well have been laid out at the same time.

Fig. 4. Staking out the Dalby church in 10 steps. Note: the cord must be at least √3 long

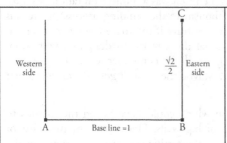

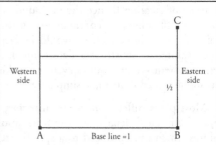

Step 1. The base line AB = 1 is staked out. It corresponds to the inside of the south wall of the south aisle. By using the cord to make a square, the eastern and western sides are staked out perpendicular to AB and northwards to the distance of √2/2. The end at the eastern side is marked with a plug C.

Step 2. Along the two sides the distance ½ is marked out and a string parallel to AB is drawn. This will be the inside of the northern arcade of the nave (which is the present northern wall of the church to day)

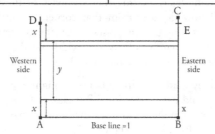

Step 3. Along the two sides the distances x and y are marked out and connected by lines parallell to AB. Thereafter the distance x is added again on top of the y-distance on both sides.

The end point of the top x on the western side is point D, and on the eastern side it is point E.

The two lines limiting the y-distance designate the outer outlines of the arcade walls. The width of the northern arcade wall has now been produced. This width goes for all the walls and is called "the width of the wall" in this text.

The x-distance equals the inner width of the south aisle.

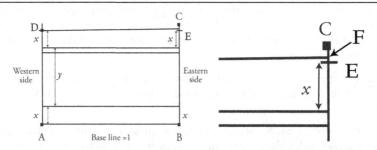

Step 4. The irregularity at the north east corner: The width of the wall is added from C and downwards the eastern side to point F, just above point E. From F a line is connected to point D on the western side, producing the interior outline of the northern wall of the north aisle.

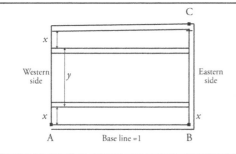

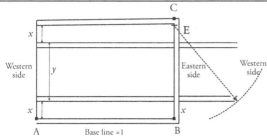

Step 5. The width of the wall is added to all the threads as parallel lines depicting the outlines of the walls. As the original location of the western wall is unknown, the outlines are not drawn here. The narthex could have been as big as a transept or as small as an aisle.

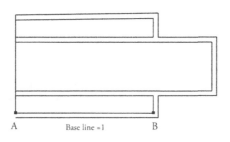

Step 6. The outlines of the arcade walls are elongated towards east. From point B and from E respectively the distance √2/2 is drawn to cut the outer elongations of the arcade walls, marking the eastern corners of the exterior of the choir.

Step 7. The outlines of the choir walls are set out.

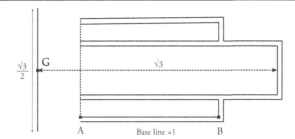

Step 8. The longitudinal axis of the basilica is staked out by adding x to y/2 upwards from A and B. From the outline of the inner east wall of the choir and westwards the distance √3 is marked out with a peg at point G, producing a total inner length of √3. Point G will be the middle point of the interior west wall of the narthex. This wall is then set out, parallell to the western side. As it is √3/2 wide, the distance √3/4 is marked out on each side of G.

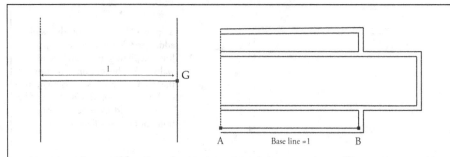

Step 9. From point G the distance 1 is set out westwards along the elongated axis. The distance corresponds to the atrium plus the width of the adjacent walls of the narthex and the royal mansion.

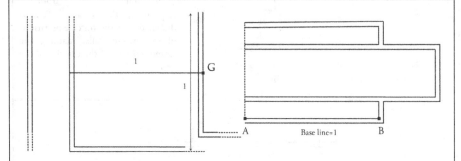

Step 10. From the north end of the narthex and southwards the distance 1 is staked out. This will be the distance to the extension of the outline of the south wing (the inside of the north wall).

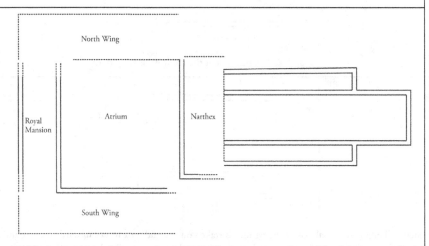

Step 11. The inner width of the mansion is 3 x 2,56 meters minus the width of the outer wall. A measuring rod of 1/10 of the module = 2,56 meters might have been used for staking out that distance. The inner width of the southern wing is uncertain, but it is most probably the same. The total length of the mansion could not be excavated, but it was estimated to be "about 26 meters". If a northern wing was erected, it was probably placed where a medieval house still stands today.

The Cord

The cord as a tool in the staking-out process can be traced considerably far back in time. At Abydos there is an inscription that tells of the sacred moment when the Egyptian Pharaoh is "stretching the cord" at the commencement of the construction of a new temple: with the help of the circumpolar stars and a water clock the Pharaoh first determined the orientation of the temple [Schneider 2002]. Then, together with Sechat, the mistress of divine books, the Pharaoh laid out the sides of the planned temple with a cord stretched between stakes. They "determine its four angles by striking the stakes with a golden mallet" [de Lubicz 1985: 94-95]. This divine act is not quite comprehensible, but perhaps it is reflected today by carpenters producing straight lines: a chalked string is held taut and then abruptly released to reveal a straight line to guide construction. When Pharaoh struck the stakes, the vibrating cord would produce an imprint on the ground. A similar technique could have been employed at the outset of the buildings in Dalby (how to determine the four angles is described by Peter Schneider [2002]).

The use of a cord as a measuring device is also mentioned in the Old Testament in Ezekiel's vision when a man, "whose appearance was like the appearance of brass" (Ezekiel 40: 3), uses a measuring cord to enlighten the prophet as to the exact measurements of the next temple (which are explained in detail in Ezekiel 40-47). The cord is actually mentioned even further back in time: when the Sumerian godess Inanna is drinking wine with Ea, she is presented with the seven pairs of *me*. The *me* is like the Egyptian *Ma'at*, the universal measures or rules for correct behavior, whether you are a star, a planet, a plant or a human being. The first *me* Inanna is given is "the stick and the cord". When she later descends to the netherworld she must let go of the seven *me* again, the last of which is the stick and the cord.

As Inanna was regarded as the Queen of Heaven, I can only interpret the items in her hands as measuring tools: they are rulers for creating the world and everything in it, and for how all these things shall move and relate to each other. The cosmic woman with the cord is also recognized as the mythical spider woman, known in the Americas as well as in Greece, and as the spinning goddess of fate, seated at the celestial pole – North or South – turning her spindle and thus making the heavens revolve. Isis and Ishtar were both depicted with spindles in their hands, and they were called "Queen of Heaven" which obviously was a function and not an empty title. It corresponds to the fact that a former pole star was called the Spinning Maiden in China and Japan [Olcott 1936: 261].[11] In a Scandinavian myth, when the hero Helge is born, a valkyrie casts out her cosmic thread into the sky to stake out the time and space of his reign. Maybe the old measures for yarn and skeins have something to tell us about measuring? In Sweden they express the ancient duodecimal system with parallels to geometry and astronomy, evidenced by the fact that numbers such as 1080 are frequently used.[12]

Perhaps it is worth looking closely at the fragments of prehistoric threads, yarn and rope previously found by archaeologists. Very often these finds are interpretated as parts of a "net" for fishing or hunting. I think that such nets were also intended for carrying food, luggage and children, but is it not also possible that some of these items might in fact have originally been measuring tools? An example of such a revised interpretation has recently been given in Denmark. Dancer Anni Brøgger borrowed all the bronze age "Egtved girl's" reconstructed equipment from the Danish National Museum for an experiment. She wore the string skirt, the blouse and the adornments every day in the summer while doing basic chores (gathering herbs, cooking, taking care of children), and demonstrated that the string skirt was extremely unpractical for any activity other than dancing [Brögger 2003: 21].[13] Besides, she realized that a long wollen thread with knots which persistently had been assumed to be a "hair net" was in fact a measuring device for making the skirt.[14] The distance between the knots directly correspond to the length of the strings and to the adjustments on the belt. The knotted thread, then, could very well have been a "recipe" for producing the skirt.

This is the same function the hypothetical cord in Dalby would have served. It was a recipe, a geometric formula, for constructing the plan. It was probably produced for this task alone and required no marks for any other units such as yards, cubits or feet.

Such a cord formula of a basilica-to-be would have been precious. Without it, the master was not able to "raise the building from the ground" which perhaps could be a new interpretation of the "*Uszuge us dem Grunde zu nemen*", as is said in the late medieval Regensburg ordinates.[15] How to do this should be kept secret. Until now, I did not understand the neccessity of keeping the builder's knowledge secret. The specialization of carpentery, of ashlar dressing and of designing demanded so many years of practice that it seems hard to imagine pirate copying by laypersons anyway. But if such a formula of knotted cord really existed, it would wisely have been kept very safe, possibly even secret.

Perhaps that is why we know so little about it.

Acknowledgments

The staking out system with the folded cord was first described briefly in "Den berömda platsen" [Mogensen 2010]. Thanks to Rheem Al-Adhami and Petter Lönegård for the illustrations. Special thanks to Rheem Al-Adhami for editing the text and assistance with translating concepts, to Erik Cinthio for important tips, and to Niels Bandholm for checking the geometry.

Notes

1. *Ipse primus dalbyensem ecclesiam edificavit*, in "*Memoriale fratrum*" [Weibull 1923: 69].
2. One of King Sven's five sons, all of whom became Danish Kings, was buried at "*locus celebris, qui Dalby, id est Vallis villa, lingua Danica dicitur*", according to Aelnoth, *Life of St. Canute the Martyr*, written in 1109; see [Cinthio 2006: 32, 88].
3. The corner of an original western part (probably of a narthex) has been overlooked by most researchers. It was however mentioned by Sten Anjou [1930: 30-31], who was in charge of the first modern excavation around 1920. It was later well documented and described by Karin Andersson in her Archaeological report to Riksantikvarieämbetet, Lunds universitets historiska museum (1970: 24).
4. This was discovered by Niels Bandholm, when he kindly controlled this setting out system. During this job he also identified some puzzling algebraical functions of the module 25.6. To construct a 1:$\sqrt{2}$ triangle in this way is in fact an ancient method of creating a "proper square" [Schneider 2002:213-214].
5. In this structure appears the *Thaleskonstruktion* mentioned by Klaus Humpert and Martin Schenk [2001: 346].
6. The original setting out of the city plan of Lund has not yet been published, but it is partly depicted in [Mogensen 2003: 12 and 49].
7. The measurements published by Sten Anjou [1930] show that the width of the church walls measured close to the foundation is 0.9 m. In the remaining building the walls tend to widen upwards to 0.95-1.0 m. The reported width of the walls of the western house is 1.0 m. In this staking out plan the width is 0.9 m. The original module was probably 25.6032 m., which would produce the width 0.90977 m.
8. [Koch 1993: 97]. He found the same unit in Ribe Cathedral, Denmark [1993: 113], and then [1993: 115] he discusses its relation to the Greek foot: $46.666 \times \sqrt{2} = 65.9966 \approx 66$, which is 2

Greek feet. The diagonal of a square ashlar with the side = 0.4667 meters is therefore 2 Greek feet.

9. This is measured from the archaeological survey where at least 2.5 meters is still lacking of the western end of the aisle. By applying the method $x + y + x$ the total distance along the western side of the square module would have been 17.7 m., which is 40 cm. less than BC. The original site of the western wall is not known, but was probably along that line or even further westwards.

10. See also [Nancy Wu 2002], particularly the articles by Marie-Thérèse Zenner and James Addiss.

11. For more myths about cosmic cords see [de Santillana and von Dechend 1970] and [Mogensen 1996].

12. The number 1080 is a half "age" of 2160 years, which is a twelfth of a precessional cycle of the equinoxes.

13. Curator Flemming Kaul of the National Museum has included this result of Anni Brögger's experience in his official presentation of the Egtved girl. See the video on the Internet: http://www.nationalmuseet.dk/sw59875.asp.

14. The knotted thread was not found in the dead girl's hair but in a separate little box together with an awl. The box had been placed beside her body in the oak coffin when she was buried about 1370 BC.

15. The word *Grund* is usually translated as "foundation". In "The Secret of the Medieval Masons," Paul Frankl presents two interesting interpretations of this quotation [Frankl 1945: 46].

References

ADAM OF BREMEN [Adamus Bremensis]. 2000. *Adam af Bremens krønike.*, Allan A. Lund, trans. Århus: Wormianum.

ANJOU, Sten. 1930. *Heliga korsets kyrka i Dalby.* Göteberg.

BRÖGGER, Anni. 2003. *Egtvedpigens dans.* Mammut.

CINTHIO, Erik. 2006. *Minnen från Lund och Dalby.* Lund: Art Factory.

FRANKL, Paul. 1945. The Secret of the Medieval Masons. *The Art Bulletin* 27, 1: 46-60.

HAAGENSEN, Erling. 1993. *Bornholms mysterium.* Copenhagen: Bogans Forlag.

HUMPERT, Klaus and Martin SCHENK. 2001. *Entdeckung der mittelalterischen Stadtplanung. Das Ende vom Mythos der "gewachsenen Stadt".* Stuttgart: Theiss.

KOCH, Mogens. 1993. *Geometri og bygningskunst.* Copenhagen: Christian Ejler.

DE LUBICZ, R. A. Schwaller. 1985. *The Egyptian Miracle: The Wisdom of the Temple.* New York: Inner Traditions.

MOGENSEN, Lone. 1996. *Himlasagor och stjärnmyter.* Stockholm: Alfabeta.

———. 2003. *Kosmisk by. ärkebiskoparnas Lund.* Malmö: Förlagshuset Nordens Grafiska Ab.

———. 2010. Den berömda platsen. In *Dalby kyrka. Om en plats i historien,* Anita Larsson, ed. Lund: Historiska Media.

OLCOTT, William Tyler. 1936. *Star Lore of All Ages.* New York: G. P. Putnam's & Sons.

DE SANTILLANA, Georgio and Herta VON DECHEND. 1970. *Hamlet's Mill.* Boston: Gambit.

SCHNEIDER, Peter. 2002. The Puzzle of the First Square in Ancient Egyptian Architecture. Pp. 207-221 in *Nexus IV: Architecture and Mathematics,* Kim Williams and Jose Francisco Rodrigues, eds. Fucecchio (Florence): Kim Williams Books.

WEIBULL, Lauritz. 1923. *Necrologium Lundense.* Lund.

WU, Nancy, ed. 2002. *Ad Quadratum. The Practical Application of Geometry in Medieval Architecture.* Aldershot, UK and Burlington VT: Ashgate.

ZENNER, Marie-Thérèse. 2002. Villard de Honnecourt and Euclidean Geometry. *Nexus Network Journal* 4, 2 (Autumn 2002): 65-78.

About the author

Lone Mogensen is a retired archaeologist and author.

Michael J. Ostwald

University of Newcastle
School of Architecture
and Built Environment
Faculty of Engineering
and Built Environment
Callaghan, New South Wales
2398 AUSTRALIA
michael.ostwald@newcastle.edu.au

Keywords: Justified plan graph,
graph theory, Space Syntax,
mathematical analysis, plan
analysis

Research

The Mathematics of Spatial Configuration: Revisiting, Revising and Critiquing Justified Plan Graph Theory

Abstract. The justified plan graph (JPG) was the first practical analytical method developed as part of the theory of Space Syntax, which purported to provide a graphical, mathematical and associated theoretical model for analysing the spatial configuration of buildings. In spite of early interest, in recent years relatively little research using this method has been published, perhaps because the JPG method is rarely explained in its totality and when it is, the descriptions are often inconsistent or unclear. Although it is now embedded in several software programs and its use may be more widespread, it is no better understood and after processing there is a marked lack of consistency in how the results are interpreted. This paper provides a historical background for the development of the JPG and a discussion of its conceptual or theoretical origins, followed by a "worked example" of the mathematics of the JPG. In combination with the results for two further cases, the paper identifies some important interpretative limits in the method and uses the examples to explain its potential use in design analysis. Finally, the paper discusses how the consistent application of this method to sets of related buildings is likely to produce a more valuable, and statistically viable, basis for future work.

Introduction

Developed primarily in the early 1980s in London, Space Syntax is the collective title given to a number of theories, tools or techniques that seek to draw connections between spatial configurations and social effects. While originally established for the study of architecture, Space Syntax theory has since been applied to the analysis of urban space and it has become one of the major analytical methods available for studying historic settlement patterns. Despite sporadic criticism of both its philosophical and mathematical foundations, and ongoing debate about its application, it remains a powerful conceptual tool for the analysis of the built environment.

The essential works of Space Syntax are the *Social Logic of Space* [Hillier and Hanson 1984], *Space is the Machine* [Hillier 1995] and *Decoding Houses and Homes* [Hanson 1998]. These three books by Bill Hillier and Julienne Hanson, along with a large number of additional papers by Alan Penn and John Peponis, define the primary conceptual framework of the theory. That is, Space Syntax promotes a conceptual shift in understanding architecture wherein "dimensional" or "geographic" thinking is rejected in favour of "relational" or "topological" reasoning. This shift, which is further elaborated in. the following sections of the present paper, relies on the process of translating architecturally defined space into a series of topological graphs that may be mathematically analysed (graph *analysis*) and then interpreted (graph *theory*) in terms of their architectural, urban, social or spatial characteristics.

Nexus Network Journal 13 (2011) 445–470   NEXUS NETWORK JOURNAL – VOL. 13, No. 2, 2011   **445**
DOI 10.1007/s00004-011-0075-3; *published online* 9 June 2011
© 2011 Kim Williams Books, Turin

While Space Syntax researchers have developed a range of analytical processes and theories, the focus of the present paper is the "justified plan graph" (JPG). This method has been known by a number of titles including "planar graphs" [March and Steadman 1971: 242] and "plan morphology" [Steadman 1983: 209], and it has been described as producing either a "justified graph" or a "justified permeability graph" [Hanson 1998: 27, 247]. Alternatively, the method is sometimes presented as producing a "plan graph", an "access graph" [Stevens 1990: 208] or a "justified access graph" [Shapiro 2005: 114]. Most recently, naming of the method has tended to return to the earlier, more general descriptors including "node analysis" and "connectivity graph analysis" [Manum 2009]. While Steadman [1983] provides different definitions for a plan graph and an access graph, the two concepts have been largely melded in subsequent use. To further complicate matters, Hillier and Hanson distinguish the syntactic analysis of urban settlements, which they call "alpha-analysis" [1984: 90] from the analysis of the interior, which they call "gamma-analysis" [1984: 147]. This means that the plan graph may also be described as a "gamma map" [1984: 147], an appellation which infers membership of a category (interior analysis) rather than a specific type of graph. For consistency, in the present paper the method is described as producing a "justified plan graph" (JPG).

The confusion surrounding the naming of this form of analysis may be one of many factors that have limited its application. Another reason is suggested by Dovey, who argues that the real problem is that "Hillier's work is at times highly difficult to understand" [1999: 24]. With its often opaque language "of 'distributed' and 'non-distributed' structures which reveal 'integration' values measured by a formula for 'relative asymmetry'" [1999: 25] the theory underlying the construction of JPGs is not easy to comprehend. Furthermore, the "evidence" for the method is typically presented "in the form of complex mathematical tables" [1999: 25] with little or no explanation of the origins of the values [Manum 2009]. Dovey is critical both of the lack of clarity of the mathematics underlying the theory and of the fact that as the opacity of the method grows, "so does the danger that such approaches may be … defended from everyday critique by their technical 'difficulty'" [1999: 25].

It is at least partially true that no single, clear and concise explanation of the current state of the JPG method exists in any one publication.[1] In a field that is still under development, albeit at a much slower pace than in the 1980s, it is not unusual to find that even the canonical works [Hillier and Hanson 1984; Hillier 1995; Hanson 1998] can no longer be considered as providing a complete and consistent explanation of JPG analysis or theory. What they do provide is a window into the state of JPG research at the time each book was written. Since then, explanations of JPG analysis have tended to be less consistent and are typically self-referential. As Klarqvist [1993] observes, they use a mix of nomenclature and often-unexplained variations of the key formulas. Conversely, some of the best descriptions of the JPG can be found in works which are either never cited as part of Space Syntax [Steadman 1983] or are critical of it [Osman and Suliman 1994]. Thus, while it is possible to construct a full and complete (if not consistent and transparent) explanation of the JPG method from published materials, it requires, as Dovey [1999] observes, an extensive investment of time and energy.

In response to this situation, the present paper provides a background to the rise of Space Syntax along with an explanation of the key theoretical concepts associated with graph theory and the JPG. Thereafter the paper is divided into two halves: the first on the mathematics of the JPG, or *graph analysis*, and the second on its visual and architectural interpretation, *graph theory*. In the first of these major sections a fully worked example of the mathematics of the JPG is offered along with a critical commentary on each stage. Inspired by Hanson's [1998] explanation of the theory, the paper proposes three similar hypothetical building plans as the focus, the first of which is the subject of the worked example (JPG analysis), while the latter two are central to the discussion of the interpretation of the results (JPG theory).

While it could be said that all of the stages of the JPG approach have been recorded in some form in previous research, the construction of a consistent mathematical and theoretical framing is the first aim of the present paper. As previously demonstrated, scholars have identified a serious need for such a description. The secondary intent of the paper is to balance the descriptive or analytical with the reflective or theoretical; it is apparent that even at its most advanced stage, there was widespread debate about the validity of the JPG method and the usefulness and meaning of its results. While the purpose of the present paper is neither to respond to past criticisms, nor to blindly repeat the standard answers of the Space Syntax movement, its commentary is informed by both of these dimensions. Finally, the paper concludes with a discussion of the need for the development and publication of consistently produced, statistically viable sets of JPG results. This process may, in turn, lead to the generation of a range of genotypes that can be used for benchmarking approaches to spatial configuration in architectural design.

Graph theory

Graph theory is conventionally regarded as originating in the seventeenth-century paradox of the Bridges of Königsberg: a mathematical puzzle about seven bridges separating four landmasses and a Knight's desire to cross each bridge only once while moving in a continuous sequence [Harary 1960; Hopkins and Wilson 2004] (fig. 1). This problem was famously solved by Euler who, in 1735, removed all of the geographic and urban complexity from the puzzle to focus only on a diagram of four nodes (landforms) and seven connections (bridges) (fig. 2). Euler used this graphical method to prove that it wasn't possible to complete the Knight's desired journey for the particular set of spatial conditions.

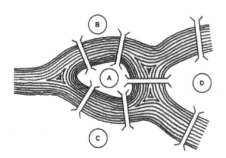

Fig. 1. The geography of the Bridges of Königsberg, based on [March and Steadman 1971]

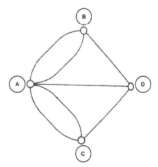

Fig. 2. A plan graph of the Bridges of Königsberg, based on [March and Steadman 1971]

While isolated examples of graph theory may be found in nineteenth-century mathematics, it wasn't until the 1960s and early 1970s that there was a growth in interest in the potential of graph theory to explain a variety of spatial and geographic phenomena [Harary 1969]. Moreover, at around this time graph theorists began to apply simple mathematical calculations to their node and line (or vertex and edge) diagrams to calculate the relative depth of these structures [Seppänen and Moore 1970; Taaffe and Gauthier 1973]. These same formulas provided the mathematical basis for Space Syntax research a decade later and in the intervening years they were responsible for encouraging a range of mathematical applications in architecture. For example, having published a simple computational model of architectural design, Christopher Alexander [1964] soon developed a variation of graph theory to explain urban connectivity [1966] before combining graph theory and a rule-based grammar to define a pattern-based approach to design [Alexander et al. 1977]. However, within a few years of this publication Alexander rejected the mathematics of graph theory, preferring instead to seek geometric or relational systems in graphs.

Similarly, March and Steadman [1971] collaboratively developed the early stages of a syntactical model of form that drew on graph theory before later separately producing extrapolations of this idea [March 1976; Steadman 1973; 1983]. However, despite being inspired by graph theory, the work of March and Steadman soon developed a focus on architectural form and graph theory was less useful for that research direction. This view was reinforced by Stiny [1975] who, along with March and Steadman [1971], could be considered as having laid the foundations for both Shape Grammar analysis and Design Computing in architecture. This meant that by the 1980s the only architectural researchers to retain a serious interest in graph theory were Hillier, Hanson and their colleagues.

Form and space

In 1984 Bill Hillier and Julienne Hanson argued that "[h]owever much we may prefer to discuss architecture in terms of visual styles, its most far-reaching practical effects are not at the level of appearances at all, but at the level of space" [1984: ix]. For Hillier and Hanson space is the fundamental medium through which architects provide shelter, structure society and serve the basic needs of communities. This idea is emphatically expressed in Hillier's adage, "space is the machine" [1995], the central maxim of Space Syntax theory. However, in order to understand Hillier and Hanson's proposition it is first necessary to place it in context.

Critics and historians have traditionally defined architecture as the art and science of constructing form. Architectural form has shape, dimensionality and actual or intended physical properties [Gelernter 1995]. The particular way in which a form is modulated or moulded, in combination with its tectonic expression and underlying structure, is a reflection of the style of the architecture [Birkerts 1994]. As a result of the focus on form, texture and materiality, historians and critics have developed elaborate techniques for classifying architecture from various eras and for hypothesising the impact these buildings have on peoples' physical and emotional responses. This focus on form as structure and as style is not surprising, given that the oldest definition of architecture, arising from Vitruvius, embraces works which exhibit suitable refinement in *firmitas*, *utilitas* and *venustas*, translated as either "soundness, utility and attractiveness" [Rowland and Howe 1999: 26] or as "firmness, commodity and delight" by Henry Wotton in 1624. Two of these three categories, firmness and delight, are directly tied to issues of form and are readily analysed and critiqued from this perspective. The third category, utility or commodity, has been less clearly defined. Vitruvius talks of *utilitas* in a design as facilitating "faultless, unimpeded use through the disposition of space" [Rowland and Howe 1999: 26]. Thus, the word "utility" suggests a degree of usefulness or functionality and the adjective "commodious" refers to things that are generous, capacious or accommodating. Both of these translations suggest a concern for spatial rather than formal properties. But what then is the relationship between space and form in architecture?

Ching describes the relationship between form and space as a "unity of opposites" [2007: 96]. He suggests that form in architecture refers to the "configuration or relative disposition of the lines or contours that delimit a figure" [2007: 34]. In contrast, space is that which is either enclosed by, or shaped by, form. Thus, a building delineates both the space it contains (its interior) and, to a lesser extent, the space in which it is contained (its site or context). Ching identifies the figure-ground plan as a rare example of a type of representation which privileges space over form: a visual device to encourage the eye to view the reciprocal relationship between solid and void. Architectural theories are conventionally focussed on form. From Pevsner's [1936] celebration of symbolic architecture to Frampton's [1995] call for a regional tectonic practice, form is central to our ethical or moral interpretation of design. Similarly, from Jencks and Baird's [1969] meditations on semiotics to Pallasmaa's [2005] phenomenology of place, architecture is read through its formal expression and the way in which the human body experiences or interprets that expression. All of these ways of viewing architecture are drawn from two of the three pillars of classical Vitruvian thought – firmness and delight. The final pillar, commodity, and especially insofar as it refers to spatial configuration, has had, in relative terms, little impact on the analysis of architectural history.

In order to overcome this deficit, Hillier and Hanson [1984] developed a theory of space *without* form. They argue that space may be empty, invisible and amorphous, but it does have two critical qualities, depreciable difference and permeability. The first of these qualities refers to the capacity to differentiate one space from any other, and the second refers to the way in which spaces are physically connected or configured. Another way of looking at the study of spatial configuration in isolation entails the rejection of two conventional geographic concerns, "the concept of location" and the "notion of distance" [Hillier and Hanson 1984: xii]. Neither of these properties or qualities is useful for understanding space as it is isolated from form. Instead, various "morphological qualities" [1984: xii], including the relationships between spaces and their relative permeability or complexity, can be studied. In practice one of the first methods adopted for the analysis of "non-geographic" space, the JPG, borrowed from the mathematics of graph theory to propose an alternative way of viewing an architectural plan. The construction of a JPG involves an inversion of the hierarchy implicit in this conventional representational schema. The JPG emphasises, at the expense of all other information, the number of spaces and the connections between them.

Developing the JPG

In Space Syntax the first step in the process is typically the production of a "convex map". The convex map serves to translate an architectural or urban plan into a diagram that reflects the configuration of selected properties of that plan. Regardless of whether researchers are interested in plan configuration, axial mapping or visual link identification, this is a necessary transition between the architectural or urban plan and the production of a graph [Turner et al. 2001]. As Hillier and Tzortzi explain,

> [s]patial layouts are first represented as a pattern of convex spaces, lines, or fields of view covering the layout (or ... some combination of them), and then calculations are made of the configurational relations between each spatial element and all, or some, others [2006: 285].

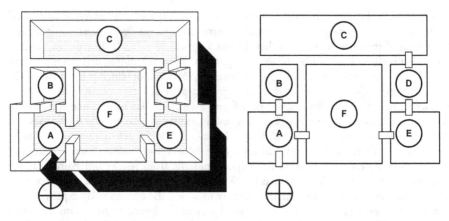

Fig. 3. Villa Alpha, cut-away view of the   Fig. 4. Villa Alpha, view of the Convex Plan
Architectural Plan

The JPG method commences with an architectural plan that records the relative size, shape, location and orientation of rooms within an overall building footprint; this could be considered the geography of the building. The connections between these rooms are simply delineated by icons for doors or openings (fig. 3). This floor plan then becomes the basis for a convex map or a representation of visible space (fig. 4).

In this case the construction of a convex map assists in the identification of spaces and connections from these architectural plans. Tom Markus explains that the construction of a convex map entails the drawing of the "fewest and fattest spaces that cover the entire plan, the former always prevailing over the latter" [1993: 14]. Earlier Hillier and Hanson had described a convex space more precisely as one wherein "no line drawn between any two points in the space goes outside the space" [1984: 98]; thus, an "L-shaped" room is a concave space and must be further broken down into two convex spaces for analysis to commence. Nevertheless, despite this advice, many subsequent authors [Marcus 1993; Hanson 1998; Dovey 1999; 2010] chose to count "L-shaped" rooms as a single node. Moreover, there are other variations that describe the delineation of a convex map by way of the social importance of parts of spaces or types of spaces, or even the relative economy of lines or zones needed to adequately describe a complete plan [Peponis et al. 1997a; 1997b]. The more recent research on convex maps is even more flexible, at least as far as the JPG variation is concerned, with Bafna [2003] foregoing the traditional convex plan in favour of using the architectural labels associated with spaces as boundary conditions.

Once the convex map is completed, a preliminary plan graph is drawn over the top of it. Conventionally at least, this graph does not differentiate between spaces that are large or small, high or low, but simply records the existence of a defined space, called a node, and whether or not the space is connected to any other space (or to the outside world). The nature of the connection – whether a door, opening, ladder, etc. – is not recorded, only the fact that a connection exists. Graphically this process converts the convex plan into a diagram of circular nodes, connected by lines (fig. 5).

Because the length of the lines is unimportant in this method – a fact that is responsible for some of the ongoing controversy surrounding the approach [Ratti 2004: Hillier and Penn 2004] – the line and node graph may then be removed from the plan and justified.

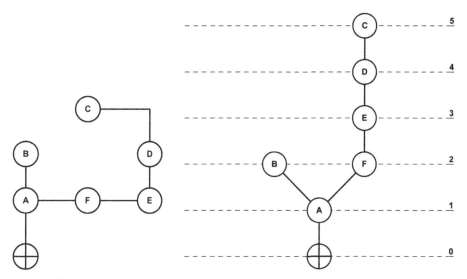

Fig. 5. Villa Alpha, view of the
Plan Graph

Fig. 6. Villa Alpha, view of Justified Plan Graph (exterior
carrier)

The word "justified" refers to the process of arranging the graph by the relative depth of nodes from a given starting point generally known as the "carrier" or "root" space [Klarqvist 1993]. Thus, the JPG is constructed around a series of horizontal, dotted lines, numbered consecutively from 0, the lowest line. Each dotted line represents a level of separation between rooms. The carrier node, often the outside world, is located on the lowest line on the chart (line 0). Those spaces that are directly connected to the carrier are located on the line above (line 1). Further spaces directly connected to those on line 1 are placed on line 2, and so on. The exterior is represented in the JPG as a crossed circle (or in text as ⊕) while other nodes are given letters (or, less often, numbers) that can be keyed to particular programmatic functions (fig. 6). Once a JPG is produced it is possible to analyse it mathematically.

JPG analysis: a worked example

While the previous section described the development of the JPG from a convex map, the present section explains the mathematics of the JPG using the same hypothetical building, the Villa Alpha, as an example. The process for mathematically analysing a building by using a JPG typically involves at least the first five steps, and possibly all nine, of the sequence that follows.

Step 1. The total number K of nodes or spaces in a set is determined. The depth of each node, relative to a carrier, is also calculated; that is, how many levels L deep in the JPG the node is. The number of nodes n_x at a given level and for a given carrier is also recorded. The number of levels is counted in the JPG from the lowest, the carrier at 0, spaces directly connected to the carrier at 1, and so on. For example, for the Villa Alpha, $K = 7$ (that is, there are 7 nodes: ⊕, A, B, C, D, E, F) and there are 6 levels (0, 1, 2, 3, 4, 5) when arrayed with ⊕ as carrier. Thus, for the JPG of the Villa Alpha with the exterior as carrier, the L value for node E = 3 (fig. 6).

Step 2. Calculate the total depth TD of the graph for a given carrier. TD is the sum of the number of connections between a particular node and every other node in the set weighted by level L. It is calculated by adding together, for each level of the justified graph, the number of nodes n_x at that level of justification multiplied by the L (0, 1, 2, 3, 4, ...). Thus:

$$TD = (0 \times n_x) + (1 \times n_x) + (2 \times n_x) + \cdots (X \times n_x).$$

For the Villa Alpha:

$$TD_{V\alpha} = (0 \times 1) + (1 \times 1) + (2 \times 2) + (1 \times 3) + (1 \times 4) + (1 \times 5)$$
$$TD_{V\alpha} = 0 + 1 + 4 + 3 + 4 + 5$$
$$TD_{V\alpha} = 17$$

This means that, with the exterior as carrier, the Villa Alpha has a $TD=17$.

Step 3. The Mean Depth MD is the average degree of depth of a node in a justified plan graph. A room depth that is higher than the mean is therefore more isolated than a room depth which is lower than the mean. MD is calculated by dividing the Total Depth TD by the number of rooms K minus one (that is, without itself). Therefore, for the Villa Alpha:

$$MD = \frac{TD}{(K-1)}$$

$$MD_{V\alpha} = \frac{17}{(7-1)}$$

$$MD_{V\alpha} = 2.833.$$

This outcome suggests that, for the Villa Alpha and with the exterior as carrier, spaces A ($L=1$), B ($L=2$) and F ($L=2$), are all more accessible than spaces C ($L=5$), D ($L=4$) and E ($L=3$).

Step 4a. Once we have TD and MD results for a given building, how can these be compared with the TD and MD results of another building, especially when it has a different number of rooms? In order to undertake such a comparison these results need to be normalised in terms of their relative depth. This outcome is called Relative Asymmetry RA, a somewhat vague term which describes an important way of normalising the range of possible results to between 0.0 and 1.0.[2] The RA for the Villa Alpha, with exterior as carrier, is calculated as follows:[3]

$$RA = \frac{2(MD-1)}{K-2}$$

$$RA_{V\alpha} = \frac{2(2.833-1)}{7-2}$$

$$RA_{V\alpha} = \frac{2 \times 1.833}{5}$$

$$RA_{V\alpha} = 0.7322.$$

When this calculation is repeated for all of the carriers for the Villa Alpha a sequence can be constructed from the most isolated node to the least isolated: "most isolated" > C (0.93), ⊕ (0.73), B (0.73), D (0.60), E (0.40), A (0.40) and F (0.33) > least isolated.

Because the *RA* results are normalised to a range between 0.0 and 1.0, *RA* results for nodes in different buildings may be usefully compared, although the veracity of this comparison is reduced if the *K* values (total number of spaces) become too dissimilar. Thus, the *RA* values of two houses, each with 9 rooms, may be directly compared. The *RA* values for two houses with, say, *K* values of 9 and 11 might also be compared, but the larger the differential between the two *K* values the less valid the comparison. In order to make a valid comparison between different size sets, an idealised benchmark must be used. For a comparative variation suitable for unequal *K* values see *Step 4b*.

Step 5a. If the *RA* for a carrier space is a reflection of its relative *isolation*, then the degree of *integration i* of that node in the JPG can be calculated by taking its reciprocal. Therefore, the integration value for the exterior space of the Villa Alpha may be calculated as follows:

$$i = \frac{1}{RA}$$

$$i_{V\alpha} = \frac{1}{0.733}$$

$$i_{V\alpha} = 1.364.$$

Once again, while this value is relatively meaningless in isolation, it is more informative when compared with either the rest of the building it is part of, or alternatively, an ideally distributed benchmark plan. In the first instance, for the Villa Alpha a comparison between *i* results for each room reveals a hierarchy of space from least integrated to most integrated as follows: least integrated < C (1.07), ⊕ (1.36), B (1.36), D (1.66), E (2.50), A (2.50) and F (3.00) < most integrated. Because of the reciprocal relationship between *i* and *RA*, this is simply the reverse order of the previous result recorded in *Step 4a*. However, whereas *RA* results were limited to a range between 0.0 and 1.0, *i* results start at 1.0 and have no upper limit. Nevertheless, in order to use this data to construct a comparison with a building of a radically different size, then a comparison must be constructed against an optimal benchmark (see *Step 5b*).

Step 4b. This is an alternative to *Step 4a* which was focused on Relative Asymmetry *RA*. Real Relative Asymmetry *RRA* describes the degree of isolation or depth of a node not only in comparison to its complete set of results but also in comparison with a suitably scaled and idealised benchmark configuration. Thus, while *RA* results are effectively *normalised* or *standardised* against a set range of results (0-1), *RRA* results are *relativised* against a benchmark configuration.

RRA results are useful for comparisons between buildings with radically different *K* values because, as buildings grow in configurational complexity and scale, their *RA* values typically fall. This outcome is a result of practical circumstances, rather than mathematical functions, but it does tend to confirm the importance of the *RRA* analytical process. Despite this, many contemporary scholars [Shapiro 2005: Thayler 2005; Manum et al. 2005] ignore *RRA* in favour of *RA* because they are not convinced by the logic of Hillier's and Hanson's method [Krüger 1989; Asami et al. 2003: Thayler 2005].

The *RRA* analytical process starts with the construction of a scalable spatial configuration against which sets of results may be relativised. The scalable configuration chosen is a diamond shape and its *RA* value is called a *D* value in recognition of this starting point. Hiller and Hanson describe the diamond configuration as one "in which there are *K* spaces at mean depth level, *K*/2 at one level above and below, *K*/4 at two levels above and below, and so on until there is one space at the shallowest [...] and deepest points [1984: 111-112]. They then provide a table of "*D* values for *K* spaces." Although rarely explained, the origins of the table may be traced to the work of Evita Periklaki and John Peponis which was published in the previous year [Peponis 1985]. The Periklaki and Peponis formula produces correct values for graphs with certain node numbers (*K* = 4, 10, 22, ...) and extrapolates for "in-between" *K* values. However, the diamond is simply one of multiple possible relativisation targets that are appropriate for architecture, and other forms, including grids, may be more appropriate for urban or communal analysis [Teklenburg et al. 1992].

The *RRA* is produced by dividing the subject *RA* by the relativised *RA* or *D* value as follows:

$$RRA = \frac{RA}{D_K}.$$

Therefore, the *RRA* for the Villa Alpha, using Hillier and Hanson's table where a *D* value for a *K* of 7=0.34, is as follows:

$$RRA_{V\alpha} = \frac{0.733}{0.34}$$

$$RRA_{V\alpha} = 2.155.$$

Step 5b. This is an alternative to *Step 5a* which was focused on Relative Asymmetry *RA*. If the *RRA* for a carrier space is a reflection of the relative isolation of a node in a JPG (in comparison with an otherwise optimal and symmetrical JPG system), then the degree of integration *i* of the JPG can be calculated by taking the reciprocal of the *RRA*:

$$i = \frac{D_K}{RA}$$

or, alternatively,

$$i = \frac{1}{RRA}.$$

The *i* value for the Villa Alpha, with exterior as carrier, and in comparison with a symmetrical configuration of the same *K* value, is therefore:

$$i_{V\alpha} = \frac{0.34}{0.733}$$

$$i_{V\alpha} = 0.463.$$

Step 6. So far, the majority of the steps have been focused on calculations for a single carrier, ⊕. At this point, steps 2, 3, 4a and 5a (or alternatively, 4b and 5b) are repeated for each other potential carrier producing a "distance data" table (table 1). In this table for the Villa Alpha, the top horizontal row of italic cells starting from 0 and ⊕ (i.e. below the column titles) record the set of results in the previous steps in this paper for the exterior carrier graph. Repeating this process for each node as carrier fills the remainder of the cells. The simplest way to do this is to produce the distance matrix in the chart, where each of the six spaces and the exterior (making seven) are placed in a matrix opposite the same space, and the number of connections needed to pass from each space to each other space is recorded. Thus, there will always be a set of cells with 0 in them where the matrix crosses. Finally, the mean results for the Villa Alpha, for *TD*, *MD*, *RA* and *i* are recorded in the table.

#	Space Vα	⊕	A	B	F	E	D	C	TD_n	MD_n	RA	i
0	⊕	*0*	*1*	*2*	*2*	*3*	*4*	*5*	17	2.83	0.73	1.36
1	A	1	0	1	1	2	3	4	12	2.00	0.40	2.50
2	B	2	1	0	2	3	4	5	17	2.83	0.73	1.36
3	F	2	1	2	0	1	2	3	11	1.83	0.33	3.00
4	E	3	2	3	1	0	1	2	12	2.00	0.40	2.50
5	D	4	3	4	2	1	0	1	15	2.50	0.60	1.66
6	C	5	4	5	3	2	1	0	20	3.33	0.93	1.07
Mean									14.85	2.47	0.59	1.92

Table 1. Distance data table for Villa Alpha

Step 7. The Control Value *CV* of a JPG is typically described as being a reflection of the degree of influence exerted by a space in a network [Jiang et al. 2000: Xinqi et al. 2008]. For example, Klarqvist describes it as "a dynamic local measure" which determines "the degree to which a space controls access to its immediate neighbours" [1993: 11]. Actually, any node has the potential to be a site of control and certain spatial configurations may increase that potential, but otherwise the *CV* has relatively little to do with power or control. The problem with the *CV* is that too often scholars have attempted to explain its value within space or architecture without first asking: What is the mathematics doing in this equation? Peponis [1985] offers one of the early definitions of the *CV* formula wherein the control value of a given point *a* is determined by the following formula:

$$CV(a) = \sum_{D(a,b)=1} \frac{1}{Val(b)}$$

In this formula *Val(b)* is the number of connections to a point, *b*.

If the *CV* equation and its operations are examined, the closest explanation for what it actually measures is offered by Asami's team who propose that control must be "thought of as a measure of relative strength ... in 'pulling' the potential [of the system] from its immediate neighbours" [2003: 48.6]. While this is reasonably close to the machinations of the formula, there is a notion in network theory entitled "distributed equilibrium" that also closely approximates the actual meaning of *CV*. Assume that a network has "capacity" of some sort and that without outside influence, this network will strive for equilibrium by automatically passing that capacity from one node equally to all adjacent nodes in the system (but no further and not back again). Once all of the capacity in the system has been simultaneously divided amongst its immediate neighbouring nodes in this way, the system will have achieved a state of equilibrium through the controlled, but unequal, distribution of its capacity. The difference between nodes in this balanced state with more or less capacity is simply a factor of adjacent network configuration. Viewed in this way the *CV* value should be thought of as signifying sites of attraction, "pulling" potential or capacity, like whirlpools that retain their position in a stream.

Jason Shapiro describes the construction of the *CV* value as beginning with "counting the number of neighbours of each space" in the JPG, that is, "the spaces with which it has a direct connection"; this is called the NC_n value. Then, "[e]ach space gives to its neighbours a value equal to 1/n of its 'control'" [Shapiro 2005: 52]. The distributed or shared value of each node is known as *CVe*: thus, $CVe = 1/NC_n$. Once the complete set of *CVe* values have been shared across the JPG, then the *CV* value for each node can be calculated. Calculating *CV* values therefore requires a holistic approach that methodically traces where every node is influenced by every connection it has. Thus, in the case of the Villa Alpha the following are three example calculations of *CV* values.

In the first example, Space ⊕ has only one connection, space A, so it must distribute 1/1 or 1 *CVe* to the space it is connected to leaving it with an interim *CV* of 0. However, space A is connected to three spaces including ⊕ and so it must distribute 1/3 or 0.33 *CVe* to each of these spaces. Thus, the *CV* for ⊕ is 0+0.33=0.33. In the second example, the *CV* for space A is calculated by taking 1/n for each of spaces ⊕ (1/1 = 1), B (1/1 = 1) and F (1/2 = 0.5). Therefore, the *CV* for space A is (1+1+0.5) = 2.50. Finally, the *CV* for space B is calculated by determining how many connections it has (NC_n=1) and placing that in the formula $CVe = 1/NC_n$, which produces a *CVe* = 1. Thus space B distributes its *CVe* of 1 to space A, leaving it with an interim *CV* value of 0. However, space A also distributes its *CVe* three ways including to B, passing a *CVe* of 0.33 back to B, giving space B a total *CV* result of 0.33.

The complete set of NC_n, *CVe* and *CV* results for the Villa Alpha are contained in table 2. They reveal that room A, with a *CV* of 2.5 is by far the space with the greatest natural attraction, followed by room D, with a *CV* = 1.5. Shapiro [2005] suggests that control values above 1.00 are considered relatively high and typically define rooms that permit or enable access. Certainly room A is a pivotal space from the point of view of access and security, but room D (the second highest) has none of these qualities. This is why the simple definition of a *CV* value as pertaining to control is less convincing than seeing it as a site of natural influence or even better of natural congregation. For Shapiro, values below 1.00 "have only weak control over adjacent spaces" [2005, 52]. If this was true, then in the Villa Alpha, nodes ⊕, B and C – all of which are terminating branches in the JPG – would have amongst the lowest capacities to exert influence over other nodes. In the case of B and C this may be true, but it is less convincing for ⊕.

#	Space Vα	⊕	A	B	F	E	D	C	NC_n	*CVe*	*CV*
0	⊕	0	1	0	0	0	0	0	1.00	1.00	0.33
1	A	1	0	1	1	0	0	0	3.00	0.33	2.50
2	B	0	1	0	0	0	0	0	1.00	1.00	0.33
3	F	0	1	0	0	1	0	0	2.00	0.50	0.83
4	E	0	0	0	1	0	1	0	2.00	0.50	1.00
5	D	0	0	0	0	1	0	1	2.00	0.50	1.50
6	C	0	0	0	0	0	1	0	1.00	1.00	0.50

Table 2. Connection Data Table for the Villa Alpha

Step 8. In this, the penultimate stage of the analytical process, the mathematics is used to provide a measure of the degree of differentiation between spaces in terms of integration. Zako [2006] observes that this stage in the analytical process has its origins in "Shannon's H-Measure" [Shannon 1949] which is a determination of transition probabilities, or entropy in information systems. A review of Shannon's formula supports this contention, although in Space Syntax the *H* measure, or Difference Factor "quantifies the spread or degree of configurational differentiation among integration values":

> [T]he closer to 0 the difference factor, the more differentiated and structured the spaces ...; the closer to 1, the more homogenised the spaces or labels, to a point where all have equal integration values and hence no configurational differences exist between them [Hanson 1998: 30-31].

It is assumed that in a set of similar projects, for example houses of the same scale, same geographic location and social structure, the distribution of space is intentional and therefore similar configurational strategies will be uncovered through the analysis of the Difference Factor. The universal solution to this question appears to be to take three values that represent the spread of results and then use this as a comparative point to test other nodes against. The spread is made up of the maximum *RA* (*a*), the mean *RA* (*b*) and the minimum *RA* (*c*). The sum of results *a*, *b* and *c*, known as *t*, is then required (i.e., *a+b+c=t*). Therefore, for the Villa Alpha *a*=0.93, *b*=0.59, *c*=0.33 and *t*=1.85 (see table 1).

The unrelativised Difference Factor (*H*) is calculated as follows:[4]

$$H = -\sum \left[\frac{a}{t} \ln\left(\frac{a}{t}\right) \right] + \left[\frac{b}{t} \ln\left(\frac{b}{t}\right) \right] + \left[\frac{c}{t} \ln\left(\frac{c}{t}\right) \right]$$

Therefore, for the Villa Alpha the calculation is:

$$H = -\sum \left[\frac{a}{t} \ln\left(\frac{a}{t}\right) \right] + \left[\frac{b}{t} \ln\left(\frac{b}{t}\right) \right] + \left[\frac{c}{t} \ln\left(\frac{c}{t}\right) \right]$$

$$H = -\sum \left[\frac{0.93}{1.85} \ln\left(\frac{0.93}{1.85}\right) \right] + \left[\frac{0.59}{1,85} \ln\left(\frac{0.59}{1.85}\right) \right] + \left[\frac{0.33}{1.85} \ln\left(\frac{0.33}{1.85}\right) \right]$$

$$H = -\sum \left[0.5027 \times -0.6877 \right] + \left[0.318 \times -1.145 \right] + \left[0.178 \times -1.725 \right]$$

$$H = -\sum \left[-0.3457 \right] + \left[-0.3644 \right] + \left[-0.3074 \right]$$

$$H = 1.0175$$

The relative difference factor, *H** normalises the unrelativised *H* result into a scale between *ln*2 and *ln*3 [Zako 2006] and is calculated as follows:

$$H* = \frac{(H - \ln 2)}{(\ln 3 - \ln 2)}.$$

For the Villa alpha this results in:

$$H* = \frac{(1.0175 - 0.693)}{(1.0986 - 0.693)}$$

$$H* = \frac{0.3245}{0.4056}$$

$$H* = 0.711.$$

While this result is calculated relative to the spread of three RA results taken from a larger set, it would be equally interesting to be able to test the relative difference factor for specific subsets or zones in a house. For example, how do the private spaces (bedrooms, bathrooms) in one house differ from those in another? Could living room nodes only be used to relativise the difference factor?

Step 9. Finally, the complete set of data for the building is tabled, recording mean, high and low results for TD, MD, RA, i and CV as well as results for H and $H*$. For the Villa Alpha, Table 3 records the complete set of results.[5]

#	Space Vα	TD_n	MD_n	RA	i	CV
0	⊕	17	2.83	0.73	1.36	0.33
1	A	12	2.00	0.40	2.50	2.50
2	B	17	2.83	0.73	1.36	0.33
3	F	11	1.83	0.33	3.00	0.83
4	E	12	2.00	0.40	2.50	1.00
5	D	15	2.50	0.60	1.66	1.50
6	C	20	3.33	0.93	1.07	0.50
Minimum		11.0	1.83	0.33	1.07	0.33
Mean		14.85	2.47	0.59	1.92	1.00
Maximum		20.00	3.33	0.93	3.00	2.50

H	1.0175
H*	0.711

Table 3. Data Summary for the Villa Alpha

Despite all of the preceding discussion about calculating TD, MD, RA and i, what most people fail to realise, and may have simply been too obvious for the Space Syntax researchers to state, is that all four of these values simply rank rooms in an identical sequence (or its inverse). These are not four ways of developing different information about a plan, they are a series of variations of the one method. For example, MD is just TD divided by the number of spaces but not counting the carrier. Thus, while the rooms will be sorted against a narrower scale, the sequence of rooms in identical. For example, in order from the highest TD to the lowest TD for the Villa Alpha the spaces are: C, ⊕/B, D, A/E, F. Similarly, in order from the highest MD to the lowest MD for the Villa Alpha the spaces are: C, ⊕/B, D, A/E, F.

Just as TD and MD are two versions of the same thing, so too are RA and i. For example, in order from the highest RA to the lowest RA for the Villa Alpha the spaces are, once again: C, ⊕/B, D, A/E, F. Because i is the reciprocal of RA, the order of rooms, from highest to lowest is simply reversed: F, A/E, D, ⊕/B, C. What is more important than the rank order using RA or i is that these results are normalised or standardised, allowing from simple comparisons between buildings. Most importantly, while beyond the scope of the present paper, in complex systems it has become increasingly important to be able to study subsets of a larger group. This could be thought of as the isolation of certain levels or distances, typically called a "radius", and this requires the type of relativisation of data that the *RA* and *i* results produce. Thus, while in one sense they do not provide any new information about the relative integration of various rooms in a plan, they have the potential, and especially in large sets, to be used for different analytical purposes [Haq 2003].

Of all of the mathematical results, the *CV* method produces something slightly different. In order from highest *CV* to lowest *CV*, for the Villa Alpha, the rooms are: A, D, E, F, C, B/⊕. This result has some similarities to the way the *i* results broadly partition spaces, but they subtly shift the balance to signify a different type of importance: pulling power rather than integration.

JPG theory: diagrammatic analysis

Having seen how the JPG analysis is undertaken mathematically, it is also possible to interpret a JPG strictly as a diagram, that is, without mathematical support [Marcus 1993; Hanson 1998; Dovey 1999; 2010]. In order to explain this type of diagramatic analysis, let us add two additional hypothetical Villas, Beta (V*β*) (fig. 7) and Gamma (V*γ*) (fig. 8) to the previously introduced Villa Alpha (V*α*).

Each of these three Villas possesses the same building footprint and the same number of rooms. The sizes of rooms and relative locations of each are also identical. Thus, from a geographic or geospatial perspective these buildings are largely identical. However, the way in which rooms are connected differs in each Villa, changing the spatial morphology in terms of permeability and depth. For example, consider a JPG, with the exterior as carrier, for each of the three villas. From this perspective it quickly becomes apparent that despite similar architectural plans, the spatial configuration present in the plan, including the degree of connectivity and depth, varies greatly (figs. 6, 9 and 10).

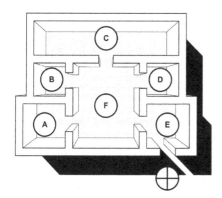
Fig. 7. Villa Beta, plan

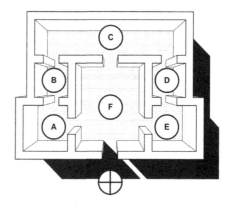
Fig. 8. Villa Gamma, plan

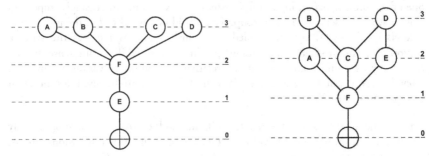

Fig. 9. Villa Beta, JPG, exterior carrier   Fig. 10. Villa Gamma, JPG, exterior carrier

From the example of the three Villas it is possible to see that while the architectural plans are similar the spatial configuration of each is quite distinctive. For example, the spatial configuration of the Villa Alpha is both linear and asymmetrical, the former implying the need to traverse various nodes along a fixed path in order to gain access to the deepest parts of the interior, and the latter suggesting a lack of spatial consistency from one side of the graph to the other. The Villa Beta, in contrast, features an arborescent ("bush-like" if it is shallow or "tree-like" if it is deeper) JPG that signifies a relatively uniform degree of spatial distribution that also implies the presence of defined zones or a hierarchical control system [Klarqvist 1993] (fig. 11). The JPG also describes a spatial configuration that is relatively symmetrical or evenly distributed from the carrier. Finally, the Villa Gamma features a more rhizomorphous structure which has been called "ring-like" in plan and which suggests a high degree of permeability or flexibility of use (fig 12).

Fig. 11. Archetypal *arborescent* visual configurations: "tree" (left) and "bush" (right)

Fig. 12. Archetypal *rhizomorphous* visual configurations: "lattice" (left) and "ring" (right)

Whereas in the Villa Alpha, room A controlled the primary rights of passage and in the Villa Beta room F fulfils a similar, if more powerful role, in the Villa Gamma multiple spaces offer alternative choices completely undermining any sense of separation or control. Finally some qualities may be discerned from a comparison between the three Villas. First, relative to houses of this size, the Villa Alpha has a very deep plan (5 layers), while the Villa Beta and the Villa Gamma both have plans of intermediate depth (3 layers). What this implies is that just by visually examining a plan graph, some qualities may be uncovered, such as relative depth / shallowness, control / permeability and symmetry / asymmetry. The first two of these categories have been used for the graphic analysis of power structures implicit in a range of institutional buildings [Marcus 1987; 1988; 1993; Dovey 1999; 2010].

While JPGs are often drawn, for ease of visual comparison, with the exterior node as carrier, a separate JPG exists for each node in a plan. One way of understanding this is to imagine that the conventional JPG, with exterior as carrier, is drawn from the point of view of a logical explorer or "visitor" who is not familiar with the building interior, but is seeking to identify and map its every space relative to the first room entered from the outside world. But a person who is familiar with the building, an "inhabitant" will use different paths to progress through the space depending on where they are located and on their familiarity with the spaces they are passing through. Thus, while the number of potential JPGs for a plan equals the number of nodes in the plan (including the exterior), this is a mathematical determination of possible configurations and people who are familiar with a building will rely on combinations of these to meet their needs. Returning to the Villa Alpha, this means that there are six remaining possible JPGs, each one generated from a different carrier room or zone (see fig. 13).

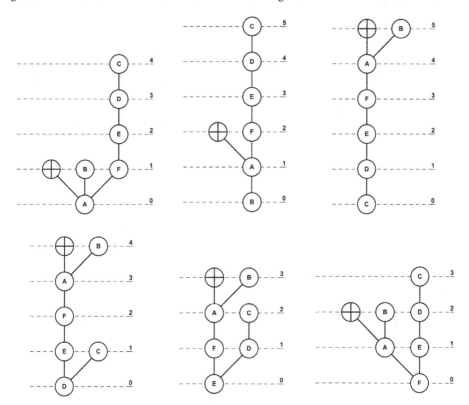

Fig. 13. Villa Alpha, the remaining six possible JPGs

While Marcus [1993] and Dovey [1999] typically only use the exterior as carrier for their visual or qualitative application of JPG theory, the reality can be much more complex. Hillier and Hanson [1984] propose that there are two types of social relations revealed in the JPG: "those between inhabitants (kinship relations or organizational hierarchies) and those between inhabitants and visitors" [Dovey 1999, 22]. Inhabitant-visitor relations can be reasonably well represented in a JPG with the exterior as carrier, but inhabitant-inhabitant relations are more complex and require consideration of multiple additional JPGs. For example a close review of the other six JPGs for the Villa Alpha shows that while, from the point of view of a visitor (carrier ⊕, fig. 6) the plan is linear and controlling, from the point of view of an inhabitant (carriers E or F, fig. 13) the plan is less linear, less deep and more symmetrical. This might suggest a somewhat defensive approach to visitors (or heightened desire for privacy), but a more balanced approach to inhabitation. The reality that multiple, parallel interpretations of the JPGs for a given building are possible, is often forgotten in graphical analysis; this fact can have significant analytical consequences. This is why, despite the usefulness of the JPG as a qualitative or visual tool, a more quantitative, mathematical approach is valuable.

JPG Theory: Interpretation and Discussion

Khadiga Osman and Mamoun Suliman argue that while the "analytical procedure of the [JPG] method is simple, objective, and replicable, the interpretation process of the numerical results remains complex, subjective and ... controversial" [1994: 190]. Edmund Leach [1978] similarly suggests that while the mathematics may be used to make simple distinctions, the bigger question remains: What does this really say about social patterns in space? In this, the penultimate section of the paper, an attempt is made to use JPG theory to interpret the analytical or mathematical results recorded in the previous section for the Villa Alpha (Table 3) along with the results for the Villas Beta (Table 4) and Gamma (Table 5). The purpose of this comparison is to test whether the mathematical results, and their standard or anticipated interpretation, are supported by a simple comparison with two different spatial configurations of the same size K.

Starting with the depth of space TD results and excluding the exterior: for the Villa Alpha the deepest space, C, has a TD of 20; for the Villa Beta the deepest spaces, A, B, C and D have a TD of 12; and for the Villa Gamma, the deepest spaces, B or D have a TD of 12. This suggests that the plan with the deepest room is Alpha and that Beta and Gamma each have multiple spaces of similar depth, although overall less than Alpha. This certainly aligns with a common sense reading of the plans but it is also apparent that Beta and Gamma, despite having the same highest level of TD, are also quite different. Thus it may be more informative to calculate the mean TD results: $TD_{V\alpha} = 14.85$, $TD_{V\beta} = 11.42$, $TD_{V\gamma} = 10.85$. Given the same K values for each of the dwellings (that is, they all have the same number of spaces) it is not surprising that the degree of difference is reduced when the average weighted depth for the spatial configuration is determined. Thus mean MD for the villas is as follows: $MD_{V\alpha} = 2.47$, $MD_{V\beta} = 1.90$, $MD_{V\gamma} = 1.80$. Thus, the plan of Villa Beta is slightly deeper on average than the plan for the Villa Gamma.

So far, all of this is in accord with the anticipated results and the standard reading of the architectural qualities of such spaces. The question of what these results are useful for is slightly more perplexing. Certainly by identifying the average or mean result for some quality in a system it is immediately possible to classify nodes into "above" or "below" average (in this case depth). While of only minimal interest in the case of the three villas, it can be a useful quantity to know for larger buildings; for example, Hanson [1998] uses the JPG method to examine Bearwood Hall (c1865), an English manor house with 134 rooms.

#	Space Vβ	TD_n	MD_n	RA	i	CV
0	⊕	15	2.50	0.60	1.66	0.50
1	F	7	1.16	0.06	15.00	4.50
2	A	12	2.00	0.40	2.50	0.20
3	B	12	2.00	0.40	2.50	0.20
4	C	12	2.00	0.40	2.50	0.20
5	D	12	2.00	0.40	2.50	0.20
6	E	10	1.66	0.26	3.75	1.20
Minimum		7.00	1.16	0.06	1.66	0.20
Mean		11.42	1.90	0.36	4.34	1.00
Maximum		15.00	2.50	0.60	15.00	4.50

H				0.766		
H*				0.181		

Table 4. Data Summary for the Villa Beta

#	Space Vγ	TD_n	MD_n	RA	i	CV
0	⊕	13	2.16	0.46	2.14	0.25
1	F	8	1.33	0.13	7.50	2.33
2	A	11	1.83	0.33	3.00	0.75
3	C	9	1.50	0.20	5.00	1.25
4	E	11	1.83	0.33	3.00	0.75
5	B	12	2.00	0.40	2.50	0.83
6	D	12	2.00	0.40	2.50	0.83
Minimum		8.00	1.33	0.13	2.14	0.25
Mean		10.85	1.80	0.32	3.66	1.00
Maximum		13.00	2.16	0.46	7.50	2.33

H				0.99		
H*				0.73		

Table 5. Data Summary for the Villa Gamma

Putting aside this problem for the present, according to the theory of Hillier and Hanson [1984], a perfect shallow and symmetrical composition should have an *RA* close to 0.00, while a perfectly linear, enfilade structure should have an *RA* closer to 1.00. The Villa Alpha is mostly linear, with one deliberate variation and its mean result, $RA_{V\alpha} = 0.59$, which is closer to a value of 1 (a linear structure) seems to confirm this.

The Villa Beta is a relatively shallow and symmetrical structure. Only the presence of the vestibule spaces, E and F, removes the distribution of nodes from an ideal structure by adding two levels of depth; this leads to a mean $RA_{V\beta}= 0.36$. This result is closer to 0 (shallow and distributed) than to 1 (linear) once again supporting, albeit in an abstract way, the standard theoretical interpretation. Finally, the Villa Gamma has the lowest mean result, $RA_{V\gamma} = 0.32$, which implies it is the most shallow of the plans, but only by a small margin.

The reciprocal of the *RA* is *i*: the integration dimension that is central to so much analysis using the JPG method [Shapiro 2005]. *i* is typically used to identify spaces, or sequences of spaces, that are pivotal to a spatial configuration and those which are not. The use of *i* to rank a set of rooms is also a special type of data set. As Sonit Bafna records, the "ranking of programmatic labelled spaces according to their mean depth (most often described in terms of integrations values)" is called an "inequality genotype" [2001: 20.1]. For the Villa Alpha, the inequality genotype, or sequence from most integrated to least integrated (or most isolated) is: F (3.00), E and A (both 2.50), D (1.66), B and ⊕ (1.36) and C (1.07). Bill Hanson suggests that this sequence is a record of "inhabitant-visitor" [1998: 29] relations and that it may be more important for a house to just consider "inhabitant-inhabitant" [1998: 29] relations. When the JPG data for the Villa Alpha is recalculated without the presence of an exterior node, then the following is the integration sequence: F and E (2.50), D and A (1.66) and B and C (1.00). This change marginally flattens the results for the Villa Alpha, identifying three clear zones of integration and replicating the visual affect of the JPG produced with F as carrier (fig. 11). The only obvious criticism of this result is that it seems to be simply identifying a space with mean depth and then extrapolates each successive reduction of i outwards from that point, a reasonable and intuitive, but otherwise simplistic result. Similar patterns of results are developed for the Villa Beta and Villa Gamma when a comparison is constructed between i for the whole set and i just for the interior set. For the former case, a range of 1.66 to 15.00 simply drops to a range of 0.20 to 5.00. In the latter case, the Villa Gamma, a range of 2.14 to 7.50 is reduced to a range of 2.50 to 5.00.

The CV results for the Villa Beta show, not surprisingly, that space F, the central hall, from which all other interior spaces connect, is not only the most influential of the spaces, but it is between 3.75 and 22.5 times more influential than any other space in the villa. While a well-informed designer would identify this space as the most important, it helps to be able to quantify this figure. For the Villa Gamma, with its ringed, permeable structure, only space C, a mid-depth, secondary foyer (and the most direct path to spaces B and D), has a slightly elevated level of influence or attraction and only the exterior node has a much-reduced level. All of the other nodes have quite similar results (a range from 0.75 to 0.25).

Before considering relative difference factors, it should be noted that it is possible for one particular spatial configuration to defy JPG analysis completely. If, in the case of the Villa Beta the major entry was directly into the central hall (F) and if all other spaces opened directly out of the hall, then a mathematical error occurs and it is impossible to calculate the $H*$ value.[6] In most other circumstances the $H*$ value is meant to offer some information about the degree to which a complete JPG is homogenous (or has similar i values) or is differentiated (has dissimilar i values).

The relative difference factor, or $H*$ value, for the Villa Alpha was 0.71 and for the Villa Gamma, 0.73: almost identical results. These results are both closer to 1.00 than to 0.00 so they fall into the category of graphs that are "more homogenised" or "where all have equal integration values" [Hanson 1998: 30-31]. Given the complete lack of similarity in the structure, depth and symmetry of these two JPGs the result is unexpected. This anomaly occurs because the range between the highest and lowest i results is quite similar in both cases. The actual i results are dissimilar, but it is the range between the highest, lowest and the mean which governs the calculation.

Just as the $H*$ values for the Villas Alpha and Gamma are not entirely useful for interpreting the difference between the two JPGs, so too the $H*$ result for the Villa Beta is unexpected: 0.181. Such a result suggests a highly differentiated or structured JPG. As a symmetrical, "bush-like" structure, with five of the rooms opening from a single common room, it is, at first, hard to reconcile this interpretation. However the central controlling hall, room F, has a very high integration value, much higher than the remainder of the rooms, supporting one particular interpretation of the idea of differentiated or structured space.

Ultimately, the use of H^* values for the analysis of three simple structures, which are already almost archetypes, is not terribly informative and a larger and more complex body of data is required to really test the interpretation of this part of the method.

Conclusion

It is difficult to support the criticism that the mathematics of the JPG method has rarely been consistently presented in its totality since it was first formulated but the task of reconstructing it remains complex, time consuming and especially involved for the general scholar. Perhaps because it was explained in so much detail in the early work of Steadman [1973: 1983], subsequent works have presented only isolated stages of the method. Similarly, perhaps because Hillier's and Hanson's major works were developed while the JPG method was still being formulated and tested, it was never given the full space it deserved. Finally, as time has passed, the major developments in JPG analysis have tended to be recorded in scattered journal and conference papers, making the task of finding, evaluating and combining works especially demanding. Once a consistent analytical method has been developed, then the problem becomes focussed on the more divisive issue of what it all means [Leach 1978: Osman and Suliman 1994: Dovey 2010].

However, despite concurring with some of the concerns raised about the mathematics and theory of the JPG, what is certain is that the larger the sample size, the more interesting the results become. For example, the studies of seventeen vernacular farmhouses in Normandy [Hillier et al. 1987], seven Pueblo "room-blocks" [Shapiro 1995] and eighteen post-war suburban houses in London [Hanson 1998] are all informative because the body of data being analysed is large enough to not only make comparisons between the various values (*TD*, *i*, *CV* etc.) but to develop hypothetical genotypes for such sets. A structural genotype is a socially authorised, ideal spatial configuration, for a particular programmatic type. It can be identified through the close analysis of multiple instances of the same type, in the same social or cultural setting. Dovey argues that the "great achievement of spatial syntax analysis has been ... [to] reveal a social ideology embedded in structural genotypes" [1999: 24]. The genotype can really only be uncovered if there is a sufficiently large set of examples being studied and thus the JPG method is most useful when applied to sets of buildings in a consistent way.

Finally, there is a dimension of JPG research that has not yet been adequately developed. While the JPG has been used to compare social patterns in a range of historic and vernacular homes or villages and in some modern housing estates, there are few examples of the method being used to analyse the body of work of an individual architect [Major and Sarris 1999; Bafna 2001]. Hanson [1998] certainly analyses some isolated architect-designed houses; she also constructs a social comparison between architect-designed and non-architect-designed houses, but this isn't the same thing. While Space Syntax and the JPG are commonly presented as being an approach to understanding social patterns embedded in a spatial configuration, they can also be used to obtain insights into the way designers think about space and into the social values they embody in their work. Thus, instead of developing a genotype for the comparative analysis of English cottages, or French farmhouses, one could be developed for Frank Lloyd Wright's prairie houses, Le Corbusier's villas or Glenn Murcutt's rural homes. There is at least one good precedent for this: Sonit Bafna's [2001] study of Miesian courtyard houses, which, while largely concentrated on methodological issues, provides an important template for future analysis. Through the focussed analysis of an architect's work it would be possible not only to interrogate their actual social and cultural values, and to compare them with their espoused views, but it may also be possible to trace the development of their design approaches over the course of their careers. It is this territory of design analysis, previously largely ignored by proponents of the JPG method, which is potentially most relevant for its revival.

Acknowledgments

I would like to acknowledge the detailed, insightful and supportive comments of Professor John Peponis. An ARC Fellowship (FT0991309) and an ARC Discovery Grant (DP1094154) supported the research undertaken in this paper. All of the images in the paper are by Romi McPherson for the author.

Appendix: Abbreviations, nomenclature and formulas

Key	Meaning	Explanation / Formula
K	Number of nodes in a set	Nodes are either: (1) spaces or rooms that are visually and spatially defined or (2) the continuous exterior of the building.
L	Level in a JPG	Not used very often, but sometimes useful for explaining the levels at which a particular node appears in a graph. Also appears in old papers as d for depth.
TD	Total Depth	The sum of the number of connections between a particular node (the carrier or root) and every other node in the set, weighted by depth. $$TD = (0 \times n_x) + (1 \times n_x) + (2 \times n_x) + \cdots (X \times n_x).$$ Because TD is relative to the carrier, it is sometimes abbreviated to TD_n which means the TD value for a particular node.
MD	Mean Depth	The average degree of depth of a node in a JPG. MD is calculated by dividing the TD by the number of nodes (K) minus one (that is, without itself). Therefore: $$MD = \frac{TD}{(K-1)}.$$ Because MD is relative to the carrier, it is sometimes abbreviated to MD_n which means the MD value for a particular node.
RA	Relative Asymmetry	A measure of how deep a system is (for a given carrier) relative to a symmetrical or balanced model of the same system. RA is calculated by the following formula: $$RA = \frac{2(MD-1)}{K-2}.$$

i	Integration Value	A measure of the degree of integration (relative centrality of spaces) in a system. i is the reciprocal of RA, thus: $$i = \frac{1}{RA}$$ Some older papers use In (uppercase I, lowercase n) as the abbreviation for integration but this has been confused with ln (lowercase l and lowercase n) which conventionally, in both mathematics and in JPG theory, means *natural log*.
RRA	Real Relative Asymmetry	Describes the degree of isolation or depth of a node not only in comparison to its own system or set, but also in comparison with a suitably scaled and idealised benchmark configuration, D. The idealised building (D) is always relative to a particular K value (a "look up" chart in Hillier and Hanson [1984] provides a value for D_K. Thereafter: $$RRA = \frac{RA}{D_K}.$$ The i value for RRA is calculated using either of the following formulas: $$i = \frac{D_K}{RA} \quad i = \frac{1}{RRA}.$$
D	Benchmark configuration for a JPG	A JPG where there are K spaces at mean depth, $K/2$ at both one level above and below that depth, $K/4$ at both two levels above and two below, and so on until there is only a single node at both the carrier and deepest levels. This results in a diamond (D) shaped "optimal" configuration.
CV	Control Value	A measure of the degree of influence each node has in a system. To calculate the CV, a determination of the NC_n and CVe of each node needs to be made as follows. NC_n (Node Connection number) is the number of spaces directly connected to a node. CVe (Control Value distributed to each node) is a redistribution of the set of relations in a JPG relative to a particular node.
H	Unrelativised difference factor	Calculated using the following formula and wherein: a = the maximum RA; b = mean RA; and c = minimum RA. The sum of results a, b and c is t (i.e., $a+b+c=t$). ln is natural logarithm to base e. $$H = -\sum \left[\frac{a}{t} \ln\left(\frac{a}{t}\right) \right] + \left[\frac{b}{t} \ln\left(\frac{b}{t}\right) \right] + \left[\frac{c}{t} \ln\left(\frac{c}{t}\right) \right].$$
H^*	Relativised difference factor	$$H^* = \frac{(H - \ln 2)}{(\ln 3 - \ln 2)}.$$

Notes

1. As part of the research undertaken for this paper a review of over fifty publications on the JPG method, spanning from the early 1970s to the present, was undertaken. This review noted in the early papers a not unexpected lack of consistency in nomenclature, interpretation and mathematical basis. Conversely, many recent papers have a tendency to rely blindly on software, a process which completely hides the mathematics of the operation and which ignores the possibility that the software might be wrong. On those occasions when the formulas were published in papers and it was possible to check the veracity of the results a surprising number of mathematical errors were uncovered.

2. In statistical analysis such a process is usually undertaken using "Z-scores". This works by standardising the complete set of results around two parameters: the mean is set at 0 and the spread of results is distributed in such a way as to attain a *standard deviation* of 1. This will produce a set of results arrayed around the 0 point with the negative results being below the original mean and the positive above it.

3. Hanson [1998: 28] offers one of the very few worked examples in the JPG field but it features a mis-transcribed figure for an *MD* result (the correct result of the *MD* calculation, 4.78, is mistyped as an *MD* value of 4.87 in the following *RA* calculation) meaning that her results are compromised.

4. Note that *ln* in the formula is natural logarithm to base *e*, not *In*, short for integration, a common error in papers using old nomenclature, or 1n, another mistake repeated in the JPG literature

5. For this paper all of the initial calculations were completed "by hand" to confirm the method and then mathematically scripted in a scientific calculator program for subsequent operations. At the completion of the worked examples, *AGraph* software (vers. 1.14) was used as a final check of results. While *AGraph* was an excellent tool for the majority of the process, it does not correctly calculate H and H^* results using natural logarithms and does not handle irrational numbers correctly.

6. A methodological problem occurs when the total depth of a node equals the number of rooms, minus one. That is, $MD = 1.00$: a room which is connected to every other room. The problem occurs because the Relative Asymmetry calculation (step 4a) produces an irrational result for the special case of a room that connects to every other room. This is a problem because the following stage (step 5a) seeks the reciprocal of the irrational number, which leads to H and H^* values being impossible to calculate for the chosen dwelling.

References

ALEXANDER, Christopher. 1964. *Notes on the synthesis of form*. Cambridge, MA: Harvard University Press.

———. 1966. A City is Not a Tree, Part I and Part II. *Design* **206** (February 1966): 46-55.

ALEXANDER, Christopher, Sara ISHIKAWA and Murray SILVERSTEIN. 1977. *A Pattern Language: towns, buildings, construction*. New York: Oxford University Press.

ASAMI, Yasushi, Ayse Sema KUBAT, Kensuke KITAGAWA and Shin-ichi IIDA. 2003. Introducing the third dimension on Space Syntax: Application on the historical Istanbul. *Proceedings: 4th International Space Syntax Symposium London*: 48.1-48.18.

BAFNA, Sonit. 2001. Geometric Intuitions of Genotypes. *Proceedings of the Third International Symposium on Space Syntax*, 20.1-20.16.

———. 2003. Space Syntax: a Brief Introduction to its Logic and Analytical Techniques. *Environment and Behavior* **35**, 1: 17-29.

BIRKERTS, Gunnar. 1994. *Process and Expression in Architectural Form*. Norman: University of Oklahoma Press.

CHING, Francis D. K. 2007. *Architecture: Form, Space and Order*. Hoboken, NJ: John Wiley and Sons.

DOVEY, Kim 2010. *Becoming Places: Urbanism / Architecture / Identity / Power*. London: Routledge.

———. 1999. *Framing Places: Mediating Power in Built Form*. London: Routledge.

FRAMPTON, Kenneth. 1995. *Studies in Tectonic Culture: The Poetics of Construction in Nineteenth and Twentieth Century Architecture.* Cambridge, MA: MIT Press.

GELERNTER, Mark. 1995. *Sources of Architectural Form: A Critical History of Western Design Theory.* New York: St. Martin's Press.

HANSON, Julienne. 1998. *Decoding Homes and Houses.* Cambridge: Cambridge University Press.

HAQ, S. 2003 Investigating the syntax line: configurational properties and cognitive correlates. *Environment and Planning B: Planning and Design* 30: 841-863.

HARARY, Frank. 1960. Some Historical and Intuitive Aspects of Graph Theory. *SIAM Review* 2, 2 (April 1960): 123-131.

———. 1969. *Graph Theory.* Reading, MA: Addison-Wesley.

HILLIER, Bill. 1995. *Space is the Machine,* Cambridge: Cambridge University Press.

HILLIER, Bill and Alan PENN. 2004. Rejoinder to Carlo Ratti. *Environment and Planning B_ Planning and Design* 31, 4: 487–499.

Hillier, Bill and Julienne HANSON. 1984. *The Social Logic of Space.* New York: Cambridge University Press.

Hillier, Bill and Kali TZORTZI. 2006. Space Syntax: The Language of Museum Space. Pp. 282-301 in *A Companion to Museum Studies,* Sharon Macdonald, ed. London: Blackwell.

HILLIER, Bill, Julienne HANSON and H. GRAHAM. 1987. Ideas are in things: an application of the space syntax method to discovering house genotypes. *Environment and Planning B: Planning and Design* 14: 363-385.

HOPKINS, Brian and Robin J. WILSON. 2004. The Truth about Königsberg. *The College Mathematics Journal* 35, 3 (May 2004): 198-207.

JENCKS, Charles and George BAIRD, eds. 1969. *Meaning in Architecture.* New York: G. Braziller.

JIANG, Bin, Christophe CLARAMUNT and Björn KLARQVIST. 2000. Integration of space syntax into GIS for modelling urban spaces. *International Journal of Applied Earth Observation and Geoinformation* 2, 3-4: 161-171.

KLARQVIST, Björn. 1993. A Space Syntax Glossary. *Nordisk Arkitekturforskning* 2: 11-12.

KRÜGER, M. 1989. *On Node and Axial Grid Maps: Distance measures and related topics,* London: University College, London.

LEACH, Edmund. 1978. Does Space Syntax Really "Constitute the Social"? Pp. 385-401 in *Social Organisation and Settlement. Contributions from Anthropology, Archaeology and Geography,* D. Green and M. Spriggs, eds. British Archaeology Reports 47. Oxford.

MAJOR, Mark David and Nicholas SARRIS. 1999. Cloak and Dagger Theory: Manifestations of the Mundane in the Space of Eight Peter Eisenman Houses. *Space Syntax: Second International Symposium, Brasilia.* 20.1-20.14.

MANUM, Bendik. 2009. AGRAPH: Complementary Software for Axial-Line Analysis. In *Proceedings of the 7th International Space Syntax Symposium.* Daniel Koch, Lars Marcus and Jesper Steen, eds. Stockholm: KTH, 2009. 070:1

MANUM, Bendik, Espen RUSTEN and Paul BENZE. 2005. AGRAPH, Software for Drawing and Calculating Space Syntax Graphs. *Proceedings of the 5th International Space Syntax Symposium,* vol. I, 97. Delft.

MARCH, Lionel, ed. 1976. *The Architecture of Form.* Cambridge: Cambridge University Press.

MARCH, Lionel and Philip STEADMAN. 1971. *The Geometry of Environment: An introduction to spatial organization in design.* London: RIBA Publications.

MARKUS, Tom. 1987. Buildings as Classifying Devices. *Environment and Planning B: Planning and Design* 14: 467-484.

———. 1988. Down to Earth. *Building Design* (July 15): 16-17.

———. 1993. *Buildings and Power.* London: Routledge.

OSMAN, Khadiga M. and Mamoun SULIMAN. 1994. The Space Syntax Methodology: Fits and Misfit. *Architecture & Behaviour* 10, 2: 189-204.

PALLASMAA, Juhani. 2005. *The eyes of the skin: architecture and the senses.* Chichester: Wiley-Academy.

PEPONIS, John. 1985. The spatial culture of factories. *Human Relations* 38 (April 1985): 357-390.

PEPONIS, John, Jean WINEMAN, Mahbub RASHID, S. KIM, and Sonit BAFNA. 1997a. On the description of shape and spatial configuration inside buildings: convex partitions and their local properties. *Environment and Planning B: Planning and Design* **24**: 761-781.

———. 1997b. On the generation of linear representations of spatial configuration. *Space Syntax, First International Symposium*, vol III, 4.1-41.18. London.

PEVSNER, Nikolaus. 1936. *Pioneers of Modern Design*. London: Faber and Faber.

RATTI, Carlo. 2004. Space syntax: some inconsistencies. *Environment and Planning B: Planning and Design* **31**, 4: 501–511.

ROWLAND, Ingrid D. and Thomas Noble HOWE, eds. 1999. *Vitruvius: Ten Books on Architecture.* Cambridge: Cambridge University Press.

SEPPÄNEN, Jouko and James M. MOORE. 1970. Facilities Planning with Graph Theory. *Management Science* **17**, 4 (December 1970): 242-253.

SHANNON, C. E. 1949. *The Mathematical Theory of Communication.* Urbana, IL: University of Illinois.

SHAPIRO, Jason. S. 2005. *A Space Syntax Analysis of Arroyo Hondo Pueblo, New Mexico.* Santa Fe: School of American Research Press.

STEADMAN, Philip J. 1973. Graph-theoretic representation of architectural arrangement. *Architectural Research and Teaching* **2**: 161-172.

———. 1983. *Architectural Morphology.* London: Pion.

STEVENS, Garry. 1990. *The Reasoning Architect: Mathematics and Science in Design.* New York: McGraw Hill.

STINY, George. 1975. *Pictorial and Formal Aspects of Shape and Shape Grammar.* Basel: Birkhäuser.

TAAFFE E. J. and H. L. GAUTHIER. 1973. *Geography of Transportation.* New Jersey: Prentice Hall.

TEKLENBERG, J. A. F. TIMMERMANS, H. J. P. WAGENBERG, A. F. 1992. Space Syntax: Standardized integration measures and some simulations. *Environment and Planning B: Planning and Design* **20**: 347-357.

THALER, Ulrich. 2005. Narrative and Syntax, new perspectives on the Late Bronze Age palace of Pylos, Greece. *Proceedings of the 5th International Space Syntax Symposium*, vol. II, 327. Delft.

TURNER, A., M. Doxa, D. O'Sullivan, A. Penn. 2001. From isovists to visibility graphs: a methodology for the analysis of architectural space. *Environment and Planning B: Planning and Design* **28**: 103-121.

XINQI, Zheng, Zhao LU, Fu MEICHEN and Wang SHUQING. 2008. Extension and Application of Space Syntax: A Case Study of Urban Traffic Network Optimizing in Beijing. *IEEE Workshop on Power Electronics and Intelligent Transportation System*: 291-295.

ZAKO, Reem. 2006. The power of the veil: Gender inequality in the domestic setting of traditional courtyard houses. Pp. 65-75 in *Courtyard Housing: Past, Present and Future*, Brian Edwards, Magdo Sibley, Mohamad Hakmi and Peter Land, eds. New York: Taylor and Francis.

About the author

Professor Michael J. Ostwald is Dean of Architecture at the University of Newcastle (Australia), Visiting Professor at RMIT University (Melbourne) and a Professorial Fellow at Victoria University Wellington (New Zealand). He has a Ph.D. in architectural history and theory and a higher doctorate (D. Sc) in the mathematics of design. He has lectured in Asia, Europe and North America and has written and published extensively on the relationship between architecture, philosophy and geometry. He has a particular interest in fractal, topographic and computational geometry and has been awarded many international research grants in this field. Michael Ostwald is a member of the editorial boards of the *Nexus Network Journal* and *Architectural Theory Review* and he is co-editor of the journal *Architectural Design Research*. He has authored more than 250 scholarly publications, including 20 books. His recent books include *The Architecture of the New Baroque* (2006), *Residue: Architecture as a Condition of Loss* (2007), *Homo Faber 1: Modelling Design* (2008), *Homo Faber 2: Modelling Ideas* (2008) and *Understanding Architectural Education in Australasia* (2009). He is also co-editor of *Museum, Gallery and Cultural Architecture* in Australia, New Zealand and the Pacific Region (2007).

Bernard Parzysz

Université d'Orléans (France)
Laboratoire André-Revuz
Université Paris-Diderot
22 avenue du Général Leclerc
92260 Fontenay-aux-Roses
FRANCE
parzysz.bernard@wanadoo.fr

Keywords: Ottoman
architecture, minarets,
polyhedra, Turkish triangles

Research

From One Polygon to Another: A Distinctive Feature of Some Ottoman Minarets

Abstract. This article is a geometer's reflection on a specificity presented by some Turkish minarets erected during the Ottoman period which gives them a recognisable appearance. The intermediate zone (*pabuç*) between the shaft and the base of these minarets, which has both a functional and an aesthetic function, is also an answer to the problem of connecting two prismatic solids having an unequal number of lateral sides (namely, two different multiples of 4). In the present case this connection was achieved by creating a polyhedron, the lateral sides of which are triangles placed head to tail. The multiple variables of the problem allowed the Ottoman architects to produce various solutions.

1 Introduction

Each religion developed specific places for worship: synagogues, temples, churches, mosques, and so forth, and the rites and ceremonies for which they were designed conditioned their architecture. Mosques are associated with Islam, and are most certainly perceived throughout the world as its most typical monuments. They accompanied very early the history of this religion, but one of their most emblematic features, the minaret, did not appear immediately and when they did, specificities occurred depending on the period and geographic and cultural area they were built in. As Auguste Choisy writes in his *Histoire de l'architecture*: "Minarets have their own geography, just as steeples have theirs". He then defines "the cylindrical type which is specific of Persia", points out that Egyptian minarets are in the shape of a "shaft with numerous balconies sticking out, its shape being no longer cylindrical but polygonal and its plan changing from one floor to another" and that "the square tower type ... is that of Tunisia, Algeria, Morocco" [Choisy 1964: II, 102, my trans.].

As a maths academic I have been developing for some years a particular interest in architecture, and especially in the ways used by architects of all times and places to link up various solids. In this article I want to focus, from the standpoint of a Western geometer and not as an historian of architecture, on a specificity shown by minarets belonging to a given geographic area (Turkey) and period (the Ottoman Empire[1]). The problem is how to connect two different volumes (as for instance a prism and a cylinder) together, aiming at producing a shape as harmonious as possible for the intermediate volume, i.e., other than just by superposing them. It is a very general and classical question for architects, which makes it necessary to bring knowledge of spatial geometry into play, especially the study of intersections of solids. It is this geometrical aspect that I would like to develop in the particular case of Ottoman minarets, because I think it gives a simple and elegant solution to the problem.

Nexus Network Journal 13 (2011) 471–486
DOI 10.1007/s00004-011-0076-2; *published online* 8 June 2011

2 Some general points about minarets

A minaret is a tower associated with a mosque, from the top of which a man (*muezzin*) calls the faithful to prayers (*adhan*). The first minarets appeared during the eighth century, nearly a century after Muhammad's death in 632: "The minaret was absent in the early mosques, and its addition was inspired by religious buildings of other religions. The main influence probably came from the churches of Syria" [Kjeilen 2010]. (On the origin of minarets, see also [Kahlaoui 2009]).

Though minarets are widely spread throughout the Islamic world, in some regions mosques generally have no minarets: "In parts of Iran, East Africa, Arabia and much of the Far East many mosques were built without them. In such places the call to prayer is either made from the courtyard of the mosque or from the roof" [Petersen 1996: headword 'minaret'].

In the greater part of the Islamic world, minarets, like mausoleums, are frequently composed of a cylindrical or prismatic shaft standing on a square base, and the connection between them may consist of a mere superimposition, as for instance in Kunya Urgench (Turkmenistan) [Chmelnitsky 2008: 71] (fig. 1).

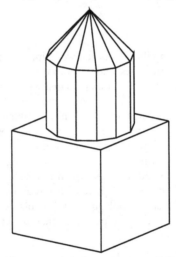

Fig. 1. Kunya Urgench. Schematic bird's eye view of Fakhr al-Din mausoleum
(end of twelfth century)

But sometimes, and relatively early, the connection of an octagonal shaft with a square base is made by cutting down the protruding trihedra, as for instance in Cairo mosques: Al-Azhar, 970 [Jacobi 1998: 72], Sultan Hassan, ca. 1360 [Meinecke-Berg 2008: 188] and Sultan Qaytbay, 1472 [Meinecke-Berg 2008: 189]. This happens in Turkey as well: the mosques of Yivli Minare (Antalya, 1230) and Ala-ed-Din (Nigde, 1223) (fig. 2).

This shape can be considered [Volwahsen 1971: 179] as resulting from the intersection of a right-angled prism and a regular pyramid, both of which have a square base (fig. 3).

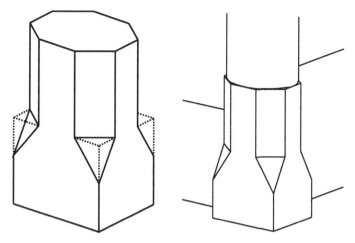

Fig. 2. a, left) Connecting an octagon to a square by cutting down trihedra;
b, right) Lower part of the minaret of Ala-ed-Din mosque (Nigde, 1223)

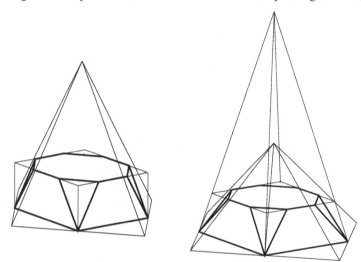

Fig. 3. Intersection of a regular prism   Fig. 4. Intersection of two regular
and pyramid with square bases   pyramids with square bases

N.B. If the octagon is inscribed in a square smaller than the base, the prism becomes a pyramid and the problem is then that of the intersection of two pyramids with square bases having the same axis (fig. 4).

Let us also note that with a dodecagonal upper part, as for instance at the Döner Kümbet[2] in Kayseri (1276), the cutting down of trihedra will be double (fig. 5).

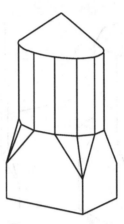

Fig. 5. Kayseri. Schematic bird's eye view of Döner Kümbet: double cutting down of trihedra

Fig. 6. Istanbul, Aghia Sofya Cami. North-west minaret

As stated by many authors, in Ottoman Turkey minarets frequently take the following form: a very slender cylindrical or prismatic shaft ending in a point and lying on a wider prismatic plinth, to which it is connected by a polyhedral intermediate part (fig. 6):

The pencil-shaped Ottoman minaret is a tall, faceted or fluted mostly polygonal or cylindrical shaft which has a ring on upper or lower part, resting on triangular buttresses (transition zone) "pabuç" above the base "kürsü". Transition zones mostly decorated with Turkish triangle motives"[3] [Altuğ 2010].

The Turkish triangle mentioned by Altuğ is "a transformation of the curved space of the traditional pendentive into a fanlike set of long and narrow triangles built at an angle from each other" (Britannica online Encyclopaedia). Like pendentives and muqarnas, the main function of Turkish triangles is to realize the junction between the polygonal plan of a building and the circular cupola topping it.

The specific shape of the minaret is possibly inherited from Egyptian Mameluke architecture, which developed in Egypt and Syria between 1250 and 1517:

> In post-Fatimid Egypt minarets developed into a complex and distinctive form. Each tower is composed of three distinct zones: a square section at the bottom, an octagonal middle section and a dome on the top. The zone of transition between each section is covered with a band of muqarnas[3] decoration [Petersen 1996: headword 'minaret'].

> In Mameluke Egypt minarets are divided into three distinct zones: a square section at the bottom, an octagonal middle part and an upper cylindrical part topped with a small cupola. ... The transition between two parts is made with a strip of muqarnas. ... During the Ottoman era, octagonal and cylindrical minarets replaced square towers. They are often high tapering minarets and, although mosques generally have a single minaret, in big towns they may have two, four or even six of them [Binous et al. 2002: 27-28, my trans.].

3 Study of a specific form of pabuç

3.1 Posing the problem

Translating Altuğ's definition into mathematical terms and observing that the lower section of Ottoman minarets is not always square (see, for instance, §§ 3.2 and 3.5 below), we can say that the base of the prismatic plinth is a convex regular polygon with $4n$ sides (n being an integer number) and the shaft, prismatic as well,[4] is a convex regular polygon with $4kn$ sides (k being an integer number, $k \geq 2$).

The study, based on examples, of a specific – though seemingly common – form of the intermediate part of Ottoman minarets, is the subject of this article. This feature is particularly noteworthy, because it poses, and gives a simple and nice solution to, the following space geometry problem (fig. 7):

> In two horizontal planes let two convex regular polygons, N and P (P being on top of N; for obvious reasons, the diameter of the circle drawn round N is bigger than the diameter of the circle drawn round P), with the same vertical axis and respectively $4n$ and $4kn$ sides ($k \geq 2$). Construct a polyhedron having the same axis, P and N being respectively its upper and lower sides.

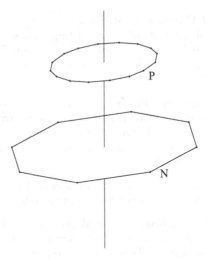

Fig. 7. ($n = 2, k = 2$)

The way in which this problem is usually solved in Turkey consists in determining plane polygons – especially triangles, which are *necessarily* plane –, some vertices of which are situated in N and the others in P. The variety of the solutions taken up by Turkish architects for determining such polyhedra will now be illustrated with some examples. (Some of the minarets studied here have been restored at various times, but nevertheless it can be assumed that the repairing did not alter their shape noticeably.) (N.B. Since actually no actual measurements were available, this study is limited to qualitative aspects.)

3.2. Example 1: Minaret of Üftade mosque, Bursa

Fig. 8. Bursa. Üftade Cami

Mehmet Muhyiddin Üftade (Bursa, 1490-1580), was one of the greatest masters of Sufism. In the Minaret of Üftade mosque in Bursa of 1579 (fig. 8) we have $n = 2$ and $k = 2$. Eight sides of the hexadecagon are parallel to a side of the lower octagon, thus creating eight isosceles trapezoids. Therefore the intermediate polyhedron, which is

---

convex, has 16 lateral sides: eight trapezoids separated by eight triangles (in the photographs the edges of the lateral sides of the intermediate polyhedron are emphasised, in order to make them more visible) (fig. 9).

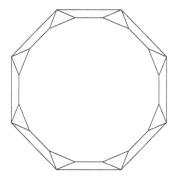

Fig. 9. Bursa, Üftade mosque. Polyhedron seen from above

This arrangement may be seen as the intersection of two regular pyramids with octagonal bases having the same axis (fig. 10).

Fig. 10. Intersection of two regular pyramids with octagonal bases

But fig. 8 shows that the arrangement of bricks visually subdivides each trapezoid into three triangles which look isosceles (fig. 11).

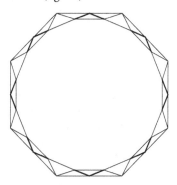

Fig. 11. Bursa, Üftade mosque. Subdivision of the trapezoids

If this is really the case, the ratio of the radii of the circles drawn around the two polygons is fixed. Actually, the length of the sides of a regular convex octagon inscribed in a circle with radius R is $A = 2R \cdot \sin(\pi/8)$ and the length of the sides of a regular convex hexadecagon inscribed in a circle with radius r is $a = 2r \cdot \sin(\pi/16)$. With the above hypothesis we have $A = 2a$. Then $R \cdot \sin(\pi/8) = 2r \cdot \sin(\pi/16)$, or $r = R \cdot \cos(\pi/16) \approx 0.98R$. The two radii are almost equal, which obviously is not the case in fig. 8. A consequence is that, among the three triangles subdividing a trapezoid, only one is isosceles.

Why the trapezoids are subdivided in such a way is also a question. I am inclined to favour a concern with aesthetics, aiming at unifying the visual aspect of the polyhedron: eventually, what is given to see is a frieze of (almost) equal triangles arranged head to tail (this is only a convenient term to indicate that the horizontal sides of two adjacent triangles are situated, one on the upper polygon, and the other on the lower polygon). This is in fact a specific use of Turkish triangles.

3.3. Example 2: North-west minaret of Aghia Sofya mosque, Istanbul

Fig. 12. Istanbul, Aghia Sofya Cami. North-west minaret

The Aghia Sofya mosque includes four minarets. The two western minarets (second half of the sixteenth century, fig. 12) are the work of the famous architect Mimar Sinan (1491-1588), who built almost two hundred buildings in Istanbul. He also built the Selimiye mosque in Edirne (1574), considered his masterpiece. Having four minarets is not a rare feature in Turkey, and the reason generally put forward is that mosques with several minarets were those built by a sultan: "In the major cities of the [Ottoman] empire mosques were built with two, four or even six minarets. At some point it seems to have been established that only a reigning sultan could erect more than one minaret per mosque" [Altuğ 2010] (see also [Petersen 1996: headword 'minaret']).

For the present minaret we have $n = 1$ and $k = 5$. It may be noticed that, as in the previous example, four sides of the upper icosagon are parallel to a side of the lower square and that, contrary to it, the vertices of the square are facing vertices of the upper polygon. The reason is that in the present case k is odd, whereas it was even in example 1. The result is that the diagonals of the square induce a "natural" connection between each side of the square and five sides of the icosagon. Then, if $k − 1$ (here 4) points are lined up at intervals along each side of the square, this will fix five successive segments transforming it into a pseudo icosagon, each side of which being associated, by a one-to-one mapping, with the nearest side of the real icosagon to form a quadrilateral (fig. 13).

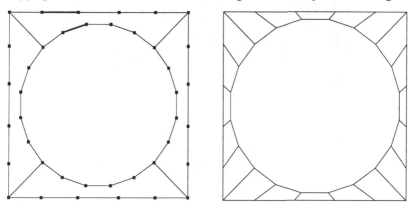

Fig. 13. Istanbul, Aghia Sofya Cami. a, left) Correspondence between the two polygons; b, right) The 20 quadrilaterals

Most of the quadrilaterals (16 out of 20) being skew, in order to get plane surfaces it will be necessary to split them into two triangles with a diagonal (fig. 14).

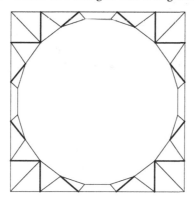

Fig. 14. Istanbul, Aghia Sofya Cami. Top view of the polyhedron

Finally, for each side of the square there will be a central trapezoid and 8 lateral triangles, four of them having their base on the square and the other four on the icosagon. Thus, on the whole there will be 4 trapezoids and 32 triangles.

Let us note that, contrary to example 1, here the trapezoids were not subdivided into triangles, doubtless because their slender shape make them similar to triangles. And the visual effect is still that of a frieze of triangles placed head to tail.

3.4. Example 3: Minaret of Aghia Sofya mosque, Iznik

Fig. 15. Iznik, Aghia Sofya Cami

In the Minaret of Aghia Sofya mosque (Iznik, fourteenth century, fig. 15) we have $n = 1$ and $k = 3$. In this case as well some sides of the dodecagon are parallel to a side of the lower square. Moreover, the vertices of the square are facing vertices of the dodecagon.

The solution adopted by the architect is basically similar to the previous one, but nevertheless it shows several differences. First the craftsman associates each vertex of the square with the vertex of the dodecagon facing it. Then, he lines up k points (i.e., 3) at intervals along each side of the square, in order to divide it into $k + 1$ consecutive sections. The segment constituted by putting together the central two sections is joined to the parallel side of the dodecagon in order to make an isosceles trapezoid (fig. 16a). This laterally determines two skew quadrilaterals which will be split into two triangles by a diagonal. To finish, contrary to the previous example but like in example 1, the trapezoids will be subdivided into three triangles, doubtless for the same reason (the division is intended to get a frieze of triangles placed head to tail) (fig. 16b). Thus a succession of 28 triangles placed head to tail is created.

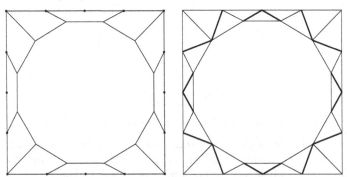

Fig. 16. Iznik, Aghia Sofya Cami. Determination of the polyhedron:
a, left) the quadrilaterals; b, right) the diagonals

In the present case this solution, already satisfying, is refined, since each of the 16 upright triangles becomes the base of a "negative" tetrahedron, created by "pushing in" a vertex under the plane of the triangle (this, so to speak, "unconvexifies" the polyhedron) (fig. 17). On the whole there is now a succession of 60 triangular sides.

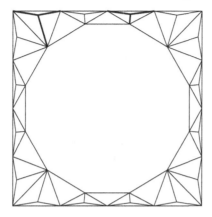

Fig. 17. Iznik, Aghia Sofya Cami. The "unconvexified" polyhedron

3.5. Example 4: West minaret of the Great Mosque of Bursa

Bursa's Great Mosque (1396, fig. 18) includes a minaret at both ends of its façade. We shall study first the West minaret.

Fig. 18. Bursa, Great Mosque (Ulu Cami). West minaret

We have here $n = 2$ and $k = 2$ as in example 1, but this time no side of the upper hexadecagon is parallel to a side of the lower octagon (thus there are no possible trapezoids). Moreover, the vertices of the octagon are facing vertices of the hexadecagon. To determine the polyhedron, the simplest solution would have been to join each vertex of the octagon with the nearest three vertices of the hexadecagon, in order to obtain 24 triangles (fig. 19a). Another solution would have been to determine k points (i.e., 2) on each side of the octagon, which would have given 40 triangles placed head to tail (fig. 19b).

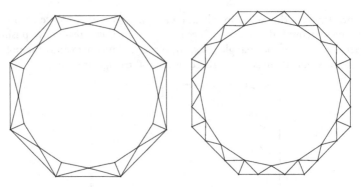

Fig. 19. Two possible connections: a, left) 24 triangles; b, right) 40 triangles

But the architect chose a third solution, certainly to get more tapering triangles (fig. 20): spread out four points on each side of the lower polygon and set a point (middle point) on each side of the upper polygon, hence 72 triangles placed head to tail.

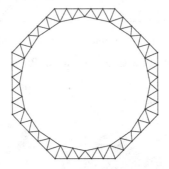

Figure 20. Bursa, Great Mosque. West minaret. The polyhedron seen from above

3.6. Example 5: East minaret of the Great Mosque of Bursa

The section of the shaft of the East minaret of the Great Mosque of Bursa (fig. 21) is hexadecagonal, as is its almost twin, but its base is square. Thus we have here $n = 1$ and $k = 4$.

Fig. 21. Bursa, Great Mosque. East minaret

As for the other minaret of this mosque, no side of the upper polygon is parallel to a side of the lower polygon. From what has been seen above, the simplest solution would have be to line up $k - 1$ points (i.e., 3) at intervals along each side of the square, in order to determine 32 triangles placed head to tail (fig. 22).

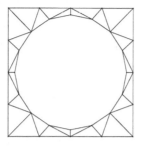

Fig. 22. A possible connection

But the triangles situated about the center of the sides of the square would then be very dissymmetric, and this is possibly the reason which led the architect to adopt another solution.

One can also see in fig. 21 that the sides of the polyhedron which are above the vertices of the square present a peculiar aspect: they are actually skew quadrilaterals made of two isosceles triangles which have the same base but are not coplanar (fig. 23).

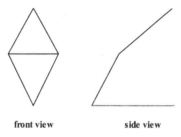

front view   side view

Fig. 23. Bursa, Great Mosque. The skew quadrilaterals in the corners

This skew quadrilateral is flanked on both sides by three triangles. The central part is constituted of five triangles, two of which (those with their horizontal side on the upper polygon) are "pushed in" to form a "negative" tetrahedron (cf. example 3). And finally we get a non-convex polyhedron with 68 lateral sides (including the sides of the "negative" tetrahedra) (fig. 24).

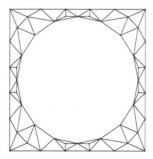

Fig. 24. Bursa, Great Mosque. East minaret. Top view

4 Conclusion

In order to synthesize the various solutions given on these examples by Ottoman architects to the question of connecting the upper part with the lower part of minarets by means of a polyhedron, it appears that seven variables can be distinguished on the whole:

- the shape of the *base* (characterised by its number of sides, $4n$);

- the shape of the *shaft* (characterised by its number of sides, $4p$);

- the integer *ratio p/n*;

- the existence (or not) of sides of the upper polygon *parallel* to sides of the lower polygon;

- the subdivision (or not) of the *trapezoids* into triangles;

- the existence (or not) of *"negative" tetrahedra* having some triangles as their bases;

- the number of *lateral sides* of the polyhedron.

This leads to Table 1, showing the great variety of the solutions found in spite of the constraints which could implicitly be observed: general structure of the minaret, number of sides of the lower polygons multiple of 4, number of sides of the upper polygon multiple of the number of sides of the lower polygon, triangles placed head to tail, and so forth.

	Example 1 Bursa Üftade	Example 2 Istanbul	Example 3 Iznik	Example 4 Bursa Ulu W.	Example 5 Bursa Ulu E.
Sides of base	8	4	4	8	4
Sides of shaft	16	20	12	16	16
Ratio	2	5	3	2	4
Parallel sides	yes	yes	yes	no	no
Subdivided trapezoids	yes	no	yes		
Negative tetrahedra	no	no	yes	no	yes
Lateral sides*	16 (32)	36	28 (60)	72	52 (68)

* For example 1, the number between brackets includes the subdivision of the trapezoids; for examples 3 and 5 it includes the sides of the "negative" tetrahedra.

Table 1. Synthesis of the studied solutions

Nevertheless, under an apparent diversity, a permanent feature can be distinguished among the minarets studied: a wish to display *visually* a frieze of triangles placed head to tail (some of them being possibly subdivided into negative tetrahedra in order to "unconvexify" the polyhedron). It resulted in a specific feature, recognizable at first sight in spite of a great variety in the arrangements of the triangles displayed between the base and the shaft. It could possibly be of interest to broaden the corpus in order to determine whether, in minarets of this type, specific values of the above variables are linked with given architects, periods or regions (throughout Ottoman empire, in Turkey and abroad[5]), but this was of course beyond the scope of this present study, which considered only the geometrical aspects of the question.

To finish with, I would just like to point out that the *pabuç* has also a functional purpose, which is to protect the minaret against violent winds and earthquakes, which occur frequently in that region. During the earthquake which shook the Izmit region of Turkey in 1999, numerous minarets collapsed or were seriously damaged. Studies undertaken subsequently [Doğangün et al. 2007] showed that the intermediate part between the shaft and the base, i.e., the *pabuç*, was the most fragile one and that Ottoman architects were apparently aware of this since, in order to reinforce this part, they had inserted iron clamps to link the blocks together; for this purpose, they used a "special technique for reinforcing and linking adjacent stone blocks with iron pieces in the vertical and horizontal directions" [Doğangün et al. 2007: 251], which proved to be effective.

Notes

1. Following Goodwin [2003], six periods can be distinguished in Ottoman architecture (1299-1922): Bursa period (1299-1437), Classical period (1437-1703), Tulip period (1703-1757), Baroque period (1757-1808), Empire period (1808-1876) and Late period (1876-1922). The minarets studied in this article belong to the Bursa and classical periods.
2. Two main types of mausoleums can be distinguished: "tower-like Seljuk graves (*türbe*) or with cupola (*kümbet*)" [Gierlichs 2008: 373, my trans.].
3. *Muqarnas*: "ornamental stalactites decorating cupolas or corbelled parts of an edifice" [Stierlin 2009: 219, my trans.].
4. When the shaft is cylindrical, the top of the intermediate part is adjusted into a convex regular polygon with $4kn$ sides, and the junction with the shaft is usually made through a torus.
5. From the mid-sixteenth to the mid-seventeenth centuries, at the height of its power, the Ottoman Empire stretched over Western Asia, Northern Africa and Southeastern Europe.

References

ALTUĞ, K. 2010. Origins and evolution of minarets. *World Bulletin* 2010/04/12. http://www.worldbulletin.net (last accessed 24 April 2011).
BINOUS, J., HAWARI, M., MARIN, M., ÖNEY, G. 2002. L'art islamique en Méditerranée. Pp 15-32 in *Genèse de l'art ottoman* (collectif). Aix-en-Provence: Edisud.
CHMELNITSKI, S. 2008. Architecture. Pp. 354-369 in *L'Islam, arts et civilisations*. M. Hattstein and P. Delius, eds. Potsdam: H.J. Ullmann.
CHOISY, A. 1964. *Histoire de l'architecture* (1899). Paris: Vincent, Fréal & Cie.
CURATOLA, G. 2009. *L'art de l'Islam*. Paris: Place des Victoires.
DOĞANGÜN, A., H. SEZEN, I. TULUK, R. LIVAOĞLU, R. ACAR. 2007. Traditional Turkish masonry monumental structures and their earthquake response. *International Journal of Architectural Heritage: Conservation, Analysis, and Restoration* I, 3: 251-271.
GIERLICHS, J. 2008. Architecture. Pp. 371-381 in *L'Islam, arts et civilisations*. M. Hattstein and P. Delius, eds. Potsdam: H.J. Ullmann.
GOODWIN, G. 2003. *A history of Ottoman architecture*. London: Thames & Hudson.
JACOBI, D. 1998. *Pascal Coste. Toutes les Égypte*. Marseille : Parenthèses.
KAHLAOUI, T. 2009. Misunderstanding the minaret. *Arab News*. http://arts-of-islam.blogspot.com (Last accessed 20 April 2011).
KJEILEN, T. 2010. Mosque. http://looklex.com/e.o/mosque.htm (Last accessed 20 April 2011.)
MEINECKE-BERG, V. 2008. Le Caire, nouveau panorama d'une capitale. Pp. 182-193 in *L'Islam, arts et civilisations*, M. Hattstein and P. Delius, eds. Potsdam: H.J. Ullmann.
PETERSEN , A. 1996. *Dictionary of Islamic architecture*. London and New York: Routledge.
SAURAT-ANFRAY, A. 2009. *Les mosquées, phares de l'Islam*. Paris: Koutoubia.

STIERLIN, H. 2005. *L'art de l'Islam en Méditerranée, d'Istanbul à Cordoue.* Paris: Gründ.

————. 2009. *Islam. De Bagdad à Cordoue. Des origines au XIIIe siècle.* Paris: Taschen.

VOLWAHSEN, A. 1971. *Inde islamique.* Collection Architecture universelle. Paris: Office du Livre.

About the author

Bernard Parzysz is Professor of mathematics – now emeritus – at the University of Orléans (France). His teaching was essentially devoted to math teachers' training and one of his research topics was – and still is – the teaching and learning of geometry. His interests include geometric mosaics, about which he published "Using Key Diagrams to Design and Construct Roman Geometric Mosaics?" in a previous issue of the *NNJ* (11, 2: 273-287).

Rachel Fletcher

113 Division St.
Great Barrington, MA
01230 USA
rfletch@bcn.net

Keywords: Thomas Jefferson,
descriptive geometry, geometric
construction, octagon, root-two

Geometer's Angle

Thomas Jefferson's Poplar Forest

Abstract. A unique geometric construction known to Thomas Jefferson reveals a rich interplay of root-two geometric elements when applied to Jefferson's octagonal plan of Poplar Forest, his eighteenth-century villa retreat.

Thomas Jefferson and classicism

In Colonial America, when buildings were typically "designed" by craftsmen and tradesmen, rather than architects, Thomas Jefferson was largely responsible for introducing the classical aesthetic to architecture. His designs reflect the neo-classical movement that emerged as Humanism in Renaissance Europe, then flourished in the Enlightenment from the 1730s to the end of the eighteenth century. Jefferson scholar Fiske Kimball considers that "directly or indirectly American classicism traces its ancestry to Jefferson, who may truly be called the father of our national architecture" [Kimball 1968, 89].

An "amateur" architect with no formal training, Jefferson first became aware of classical architecture through books, then later gained first-hand experiences of ancient Roman and eighteenth-century French buildings while serving as American Minister to Paris (1784-1789). He studied the written treatises of Marcus Pollio Vitruvius, Leon Battista Alberti, Inigo Jones, Sebastiano Serlio and others who relied on classical rules of architecture and mathematical techniques for achieving proportion [O'Neal 1978, 2]. On architectural matters, Jefferson is reported to have said that Andrea Palladio "was the bible," even though he knew his buildings only through books.[1]

Jefferson practiced the Roman classical architecture of Palladio and late eighteenth century France, and borrowed extensively from classical sources. He based the Rotunda of the University of Virginia in Charlottesville on measured drawings of the Pantheon published in Giacomo Leoni's *The Architecture of A. Palladio*. Models for the Virginia State Capitol in Richmond, which he designed with the assistance of French architect and antiquarian Charles-Louis Clerisseau, included the Temple of Balbec, the Erechtheum in Athens, and the ancient Roman temple Maison Carrée at Nîmes in France.[2] This was the first government building designed for a modern republic, the first American work in the Classical Revival style, and the first modern public building in the world to adapt the classical temple form for its exterior [Nichols 1976, 169-170; Kimball 1968, 42].

Geometric proportion

That Jefferson followed classical rules for applying simple proportions involving whole numbers to the orders and other components of building design is well documented.[3] But he also was familiar with techniques for achieving harmony through incommensurable ratios associated with elementary geometric figures. His designs often feature fundamental geometric shapes and volumes. The University Rotunda, whose dome he describes as a "sphere within a cylinder," presents an array of circles, squares and triangles.[4] For the Virginia Capitol, he selected as sources "the most perfect examples of cubic architecture, as the Pantheon of Rome is of the spherical...."[5] The plans for his Monticello residence present regular octagons, semi-octagons and elongated octagons.

Nexus Network Journal 13 (2011) 487–498   NEXUS NETWORK JOURNAL – VOL. 13, No. 2, 2011   **487**
DOI 10.1007/s00004-011-0077-1; *published online* 22 June 2011
© 2011 Kim Williams Books, Turin

There is some evidence that Jefferson applied incommensurable proportions through geometric techniques. For the Washington Capitol he specified that neighboring properties "be sold out in breadths of fifty feet; their depths to extend to the diagonal of the square." In other words, the lots conform to the incommensurable ratio 1: √2 ("Opinion on Capitol," 29 November 1790 [Ford 1904-1905, VI: 49]). His plan for the University Rotunda expresses root-two, root-three, and perhaps even Golden Mean symmetries [Fletcher 2003].

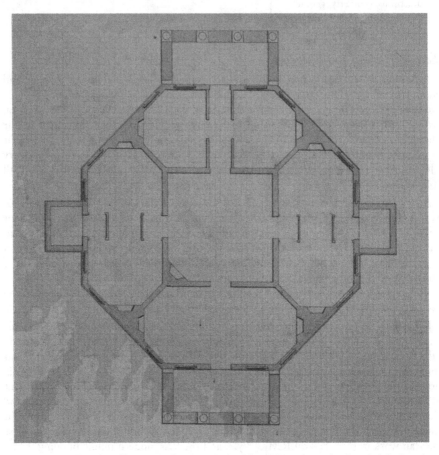

Fig. 1. Thomas Jefferson, Poplar Forest, First Floor Plan. Image (ca. 1820), inked, shaded and tinted by John Neilson (atrributed). Scale: about 10' = 1". On heavy paper, not watermarked, with co-coordinate lines drawn by hand, 9" x 11.5". (N-350, K-Pl.14). Courtesy of The Jefferson Papers of the University of Virginia, Special Collections, University of Virginia Library, Charlottesville, Virginia

Poplar Forest

A clear example of incommensurable proportion is Jefferson's octagonal villa retreat at Poplar Forest, located on the eastern slope of the Blue Ridge Mountains in Bedford County, Virginia, ninety miles southwest of Monticello, his principle residence. Building construction began in 1806, while Jefferson resided as president in Washington, supervising the project remotely through written instructions, working drawings and sketches. The villa was made habitable by the time of his retirement in 1809, but would require another sixteen years to complete [McDonald 2000, 178-81]. Jefferson first proposed the octagonal house for his Pantops farm, north of Monticello, as a future residence for his grandson Francis Eppes. But instead he realized the plan at Poplar Forest, one of his working plantations where Eppes eventually settled.[6]

Poplar Forest is of brick construction and octagonal in plan, containing a central square space, flanked on three sides by elongated octagonal rooms. On the fourth side, a short entry hall divides a pair of smaller rooms. The central dining room is skylit and measures 20' x 20' x 20', a perfect cube. On the east and west, alcove beds divide the two main bedrooms into sections [Chambers 1993, 33-35]. During construction, Jefferson added pedimented porticoes on low arcades attached on the northern and southern facades, and stairwells east and west.[7] Fig. 1, attributed to Jefferson's workman John Neilson, shows the first floor plan complete with additions to the original design.[8]

Following a fire in 1945, the interior was rebuilt leaving only the walls, chimneys and columns original. Since 1983, under the leadership of Director of Architectural Restoration Travis McDonald, Poplar Forest has been the subject of extensive research and restoration. The goal is to enable the public to experience Jefferson's retreat according to our best understanding of his original design.

Jefferson's regard for octagonal plans by Robert Morris, Inigo Jones, Palladio and others has been reviewed extensively. C. Allan Brown observes octagonal symmetry in the landscape design at Poplar Forest. And in the building plan, E. Kurt Albaugh cites the "two-fifths" rule of proportion that closely approximates the octagon's inherent root-two ratio [Albaugh 1987, 74-77; Brown 1990, 119-121; Chambers 1993, 33; Lancaster 1951, 9-10]. In fact, a specific technique for constructing the octagon, drawn more than once by Jefferson, offers compelling evidence of a consistent geometric approach to Poplar Forest's design.

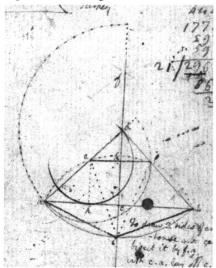 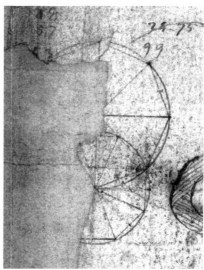

Fig. 2. Jefferson's sketches of octagons. a, left) dividing two sides of a square in root-two ratios; b, right) two completed octagons and the algebraic proof.
Images: Objects; assorted sketches and calculations, 1 page, undated, MHi29.
Original manuscript from the Coolidge Collection of Thomas Jefferson
Manuscripts. Massachusetts Historical Society [Jefferson 2003: MHi29]

A page of Jefferson's notes and scribbles, possibly executed during a visit to Poplar Forest, includes a technique for drawing three sides of a regular octagon together with two sides of a larger octagon. Both are accomplished by dividing two sides of a square, or the sides of a 45° right triangle, in root-two ratio (fig. 2a). A drawing of two completed octagons (fig. 2b) and an algebraic proof accompany the construction.[9]

How to draw three sides of an octagon on a given base

- Draw a horizontal line AB.
- Place the compass point at A. Draw a semi-circle of radius AB that intersects the extension of line BA at point D.
- From point A, draw a line perpendicular to line BD that intersects the semi-circle at point C (fig. 3).

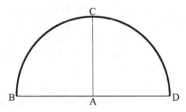

Fig. 3

- Connect points B, C and D.
- From point C, draw a circle of radius CB that intersects point D and the extension of line CA at point E (fig. 4).

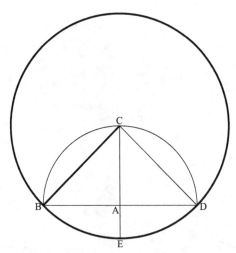

Fig. 4

- Connect points B, E and D.

The result is two sides of a regular octagon inscribed within the circle of radius CB.

- Complete the octagon (fig. 5).

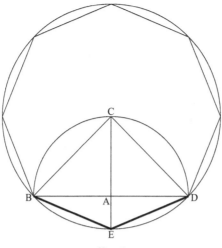

Fig. 5

- From point E, draw a circle of radius EB that intersects line BC at point F and line CD at point G.
- Connects points B, F, G and D.

The result is three sides of a regular octagon inscribed within the circle of radius EB.

- Complete the octagon.

If radius CB of the large circle is 1, side BE of its inscribed octagon equals $\sqrt{(2-\sqrt{2})}$ (0.7653...). Radius EB of the small circle therefore equals $\sqrt{(2-\sqrt{2})}$ and side BF of its inscribed octagon equals $(2-\sqrt{2})$ (0.5857...) (fig. 6).

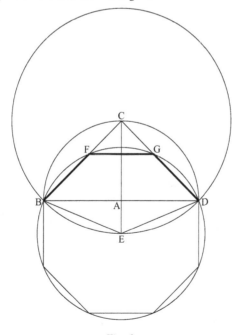

Fig. 6

We see from fig. 2a that Jefferson knew how to proceed further with this geometric construction.

- From point F, draw an arc of radius FC that intersects and is tangent to line BA at point H.
- Connect points F and H.
- Extend line FH until it intersects line BE at point I
- Alternatively from point F, draw an arc of radius FG to point I.
- Connect points G, F and I.

The result is two sides of a square.

- Complete the square (fig. 7).

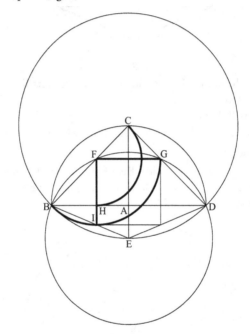

Fig. 7

Root-two symmetry

In fig. 8, the construction is scaled to Jefferson's plan for Poplar Forest. The small circle of radius EB circumscribes the octagonal footprint. Each side of its inscribed octagon, such as BF, locates an exterior wall. Center point C of the large circle locates the midpoint of the portico front.

- Repeat the semi-circle of radius CE at each quadrant, as shown.

The semi-circles intersect at points J, K, L and M.

- Complete the regular octagon.

The octagon divides into a center square, four root-two rectangles and four 45° right triangles. The square locates the center room of the Poplar Forest plan (fig. 9).

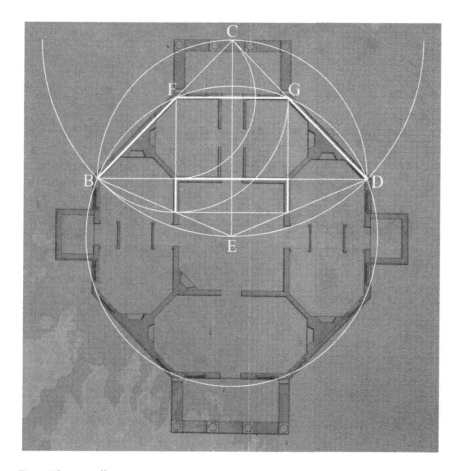

Fig. 8. Thomas Jefferson. Poplar Forest, First Floor Plan, with geometric overlay by the author

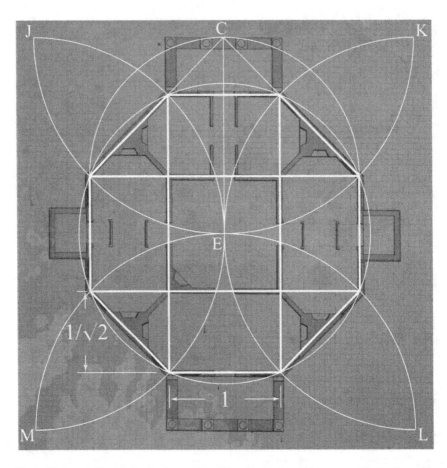

Fig. 9. Thomas Jefferson. Poplar Forest, First Floor Plan, with geometric overlay by the author

To construct the elongated octagonal rooms at each quadrant:

- Divide each 45° right triangle in half, into smaller 45° right triangles.
- Join a small 45° right triangle to each end of a root-two rectangle.

The result is an elongated hexagon (fig. 10, right).

- Locate the two sides of each small 45° right triangle. Connect their midpoints, as shown (fig. 10, upper left).
- Join the new shape to each end of a root-two rectangle.

The result is an elongated octagon that locates the room at each quadrant. Chimneys occupy the remaining spaces (fig. 10, left).[10]

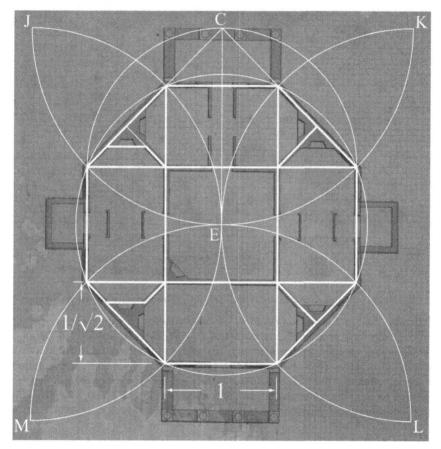

Fig. 10. Thomas Jefferson. Poplar Forest, First Floor Plan, with geometric overlay by the author

- Connect points J, K, L and M.

The result is a square that encloses the northern and southern porticoes.

- Inscribe a circle within the square.
- Inscribe a regular octagon within the circle.
- Two edges of the octagon locate the stairwells east and west (fig. 11).

Fig. 11. Thomas Jefferson. Poplar Forest, First Floor Plan, with geometric overlay by the author

Jefferson was proficient in a variety of mathematical disciplines that included arithmetic, algebra, geometry, trigonometry and Newtonian calculus, as well as mechanical and natural applications such as navigation, surveying, astronomy and geography. Geometry, which he explored in both planar and spherical configurations, held special interest. But rather than study mathematics for its own sake, Jefferson endeavored to apply his knowledge in tangible ways, as at Poplar Forest, where a simple geometric construction yields a rich, harmonic composition.

Notes

1. Colonel Isaac A. Coles to General John Cocke, 23 February 1816 [Adams 1976, 283]. Jefferson toured the agriculture of southern France and northern Italy in 1787, intending to visit Palladio's hometown of Vicenza at a later date [Nichols 1976, 163, 167].
2. Jefferson, "An Account of the Capitol in Virginia," no date, Miscellaneous Papers [Lipscomb and Bergh 1905-06, XVII: 353].
3. See [Kimball 1968]. Joseph Lasala has analyzed the Pavilions of Jefferson's University of Virginia according to the Palladian system of dividing a module, based on the lower diameter of a column, into minutes and seconds. From this are derived an order's six major components: the base, shaft and capital of the column; and the architrave, frieze and cornice of the

entablature. The order, once determined, fixes the size and distribution of other building components [Lasala 1992].

4. Jefferson to William Short, 24 November 1821 [Jefferson 2007].
5. Jefferson, "An Account of the Capitol in Virginia," no date, Miscellaneous Papers [Lipscomb and Bergh 1905-06, XVII: 353].
6. Jefferson to John Wayles Eppes, 30 June 1820 [Jefferson 2007]. Compare Jefferson's plan for the house at Pantops, before 1804, and the first floor plan of Poplar Forest drawn by John Neilson [Chambers 1993, 33-34, Kimball 1968, 182-183, figs. 193,194].
7. Jefferson proposed the additions in a letter to bricklayer Hugh Chisolm following a visit to Poplar Forest in 1806. Jefferson to Hugh Chisolm, 7 September 1806, Massachusetts Historical Society [Chambers 1993, 36-37, Kimball 1968, 182-183].
8. Although presently attributed to Neilson, Kimball in 1968 credited the drawing to Jefferson's granddaughter Cornelia J. Randolph [Kimball 1968, 183].
9. See [Chambers 1993, 21] on connecting the notes to a trip to Poplar Forest in 1800. The construction appears elsewhere in Jefferson's notebooks during various phases of building and remodeling at Monticello. One example is a theorem for drawing "three sides of an octagon" on a given base, dated 1771(?), apparently in preparation for octagonal projections in the plan. See [Jefferson 2003: N123; K94]. A similar construction, dated 1794-1795(?), accompanies studies for Monticello's remodeling. See [Jefferson 2003: N138; K140].
10. Jefferson's preliminary sketches for Pantops/Poplar Forest, before 1804, produce octagonal rooms in this fashion. See [Jefferson 2003: N260; K193].

References

ADAMS, William Howard, ed. 1976. *The Eye of Jefferson*. Washington: National Gallery of Art.

ALBAUGH, E. Kurt. 1987. Thomas Jefferson's Poplar Forest: Symmetry and Proportionality in a Palladian Summer House. *Fine Homebuilding* (October/November 1987): 74-79.

BROWN, C. Allan. 1990. Thomas Jefferson's Poplar Forest: The Mathematics of an Ideal Villa. *Journal of Garden History* 10 (April/June 1990): 117-139.

CHAMBERS, S. Allen. 1993. *Poplar Forest and Thomas Jefferson*. Forest, VA: The Corporation for Jefferson's Poplar Forest.

FLETCHER, Rachel. 2003. An American Vision of Harmony: Geometric Proportions in Thomas Jefferson's Rotunda at the University of Virginia. *Nexus Network Journal* 5, 2 (Autumn 2003): 7-47.

FORD, Paul Leicester, ed. 1904-05. *The Works of Thomas Jefferson in Twelve Volumes*. Federal Edition. New York: G. P. Putnam's Sons.

JEFFERSON, Thomas. 2003. *Thomas Jefferson Papers: An Electronic Archive*. Boston: Massachusetts Historical Society. http://www.thomasjeffersonpapers.org/

———. 2007. *Items from Special Collections at the University of Virginia Library*. Electronic Text Center. Charlottesville: University of Virginia Library. http://etext.lib.virginia.edu/speccol.html

KIMBALL, Fiske. 1968. *Thomas Jefferson: Architect* (1916). Rpt. Boston and New York: Da Capo Press.

LANCASTER, Clay. 1951. Jefferson's Architectural Indebtedness to Robert Morris. *Journal of the Society of Architectural Historians* 10, 1 (March 1951): 3-10.

LASALA, Joseph Michael. 1988. Comparative Analysis: Thomas Jefferson's Rotunda and the Pantheon in Rome. *Virginia Studio Record* 1, 2 (Fall 1988): 84-87.

———. 1992. Thomas Jefferson's Designs for the University of Virginia. Master Thesis, University of Virginia.

LIPSCOMB, Andrew A. and Albert Ellery BERGH, eds. 1905-06. *The Writings of Thomas Jefferson*. Washington, D. C.: The Thomas Jefferson Memorial Association. (Electronic version: H-BAR Enterprises, 1996.)

MCDONALD, Travis. C., Jr. 2000. Constructing Optimism: Thomas Jefferson's Poplar Forest. Pp. 176-200 in *People, Power, Places* (Perspectives in Vernacular Architecture 8), Sally McMurry and Annmarie Adams, eds. Knoxville: University of Tennessee Press.

NICHOLS, Frederick Doveton. 1961. *Thomas Jefferson's Architectural Drawings*. Boston: Massachusetts Historical Society, and Charlottesville: Thomas Jefferson Memorial Foundation and The University Press of Virginia.

————. 1976. Jefferson: The Making of an Architect. Pp.160-185 in *Jefferson and the Arts: an Extended View*, William Howard Adams, ed. Washington D.C.: National Gallery of Art.

O'NEAL, William Bainter, ed. 1978. *Jefferson's Fine Arts Library: His Selections for the University of Virginia Together with His Own Architectural Books*. Charlottesville: The University Press of Virginia.

About the geometer

Rachel Fletcher is a geometer and teacher of geometry and proportion to design practitioners. With degrees from Hofstra University, SUNY Albany and Humboldt State University, she was the creator/curator of the museum exhibits "Infinite Measure," "Design by Nature" and "Harmony by Design: The Golden Mean" and author of the exhibit catalogs. She is an adjunct professor at the New York School of Interior Design. She is founding director of the Housatonic River Walk in Great Barrington, Massachusetts, co-director of the Upper Housatonic Valley African American Heritage Trail, and a director of Friends of the W. E. B. Du Bois Boyhood Homesite. She has been a contributing editor to the *Nexus Network Journal* since 2005.

Book Review

Kim Williams, Stephen R. Wassell, Lionel March (eds.)

The Mathematical Works of Leon Battista Alberti

Basel: Birkhäuser, 2010

Reviewed by Sylvie Duvernoy

Via Benozzo Gozzoli, 26
50124 Florence ITALY
syld@kimwilliamsbooks.com

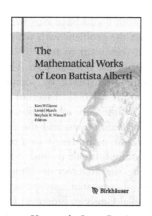

Keywords: Leon Battista Alberti, Ludi matematici, cipher codes, measuring, history of mathematics

Leon Battista Aberti spent thirty-two years in the Papal Court, earning his living with his pen. Besides being an "Abbreviator" at the Vatican, he was a prolific writer; his first personal success was a comedy, *Philodoxeos*, which he wrote at the age of twenty. His multifaceted interests led him to study different scientific disciplines and consequently the corpus of his literary production covers a wide range of topics.

The present book is dedicated to his mathematical works, and gathers four writings which all are related to mathematics; three of them also deal with another discipline. The first text, *Ex ludis rerum mathematicarum* (more often cited as *Ludi matematici*), is related to problems of survey and measurements; the second, *Elementi di pittura* (*Elements of Painting*, not to be confused with the more famous *De pictura*), is related to painting; the third, *De componendis cifris* (*On Writing in Ciphers*), is related to cryptology; the last, *De lunularum quadratura* (*On Squaring the Lune*), is the only text to deal with pure theoretical mathematics. The book proposes the transcription of a manuscript of *Ludi matematici* held in the Biblioteca Nazionale in Florence (ms. no. Galileana 10), together with a fresh translation and a commentary for each of the four texts.

The three authors of the volume are scholars in mathematics and architecture, certainly well-known to the readership of the *Nexus Network Journal* (of which Kim Williams is editor-in-chief), who have long inquired into the historical relationship between both disciplines. Prof. Stephen Wassell (mathematician) and Kim Williams (architect) have already published the translation of the short Renaissance treatise by Silvio Belli, *On Ratio and Proportion* [Belli 2003], for which Lionel March wrote a foreword. Prof. Lionel March (architect) has also written a number of important books, among which *Architectonics of Humanism: Essays on Number in Architecture* [1998] which testify to his profound knowledge in both architecture and mathematics.

Of the four mathematical writings by Alberti published here, the most important one is indeed the first, *Ludi matematici*.

Any scholar in art history and architecture history is familiar with the major writings of Leon Battista Alberti dealing with the fine arts, which comprise the famous treatise *De*

DOI 10.1007/s00004-011-0078-0; *published online* 16 June 2011
© 2011 Kim Williams Books, Turin

re aedificatoria and the book whose Latin title is *De pictura*. Both books have long been translated in English and a number of other languages. In fact the problem with which the non-Italian reading Alberti scholars are faced with, when they want to refer to his work, is: which translation to choose in order to get the most of the original text? The most recent English edition of the *De re aedificatoria* is the translation by Joseph Rykwert, Neil Leach and Robert Tavernor [1988; see also Tavernor 1998]. Tavernor is the author of the preface of this present work. Any serious architecture school library owns at least two or three different editions of both the *De re aedificatoria* and the *De pictura*. On the other hand, the booklet whose title is commonly shortened in *Ludi matematici* (literally, mathematical games or amusements) is more difficult to find. The problem is: which library should own a copy? Should it be the architecture library because the book deals with some basic historical problems of measured surveys, and because anyway Alberti was the architect of the powerful Rucellai family in Florence? Or should it be the math library because the book deals with history of applied mathematics? In Florence, the city in whose National Library the original manuscript (Galileana 10) freshly transcribed here is kept, the famous writings by Alberti are present in the various colleges' libraries, but students can only find a single copy of the *Ludi matematici* at the philosophy department of the College of Humanities. Until now, the text has always been part of a collection of writings in vulgar and has therefore been catalogued by editors – and librarians – not for its content but for its form, considering the language in which it was written: not Latin but Italian.

It seems therefore that the *Ludi matematici* has so far been much less well-known than the other writings by Alberti. In fact this edition offers the very first translation in English of the *Ludi* (more than 550 years after the manuscript's first completion) and hopefully this will make them more accessible and more desirable among scholars and students. In the introduction of Williams, March and Wassell's volume we can read:

> Both *Ludi Matematici* and *Elements of Painting* have received a substantial amount of attention in other languages, most notably Italian and French, but they have not in English (p. 1).

The commentary by Stephen Wassell enhances the importance of this somewhat underestimated text, which testifies to the cultural continuity and transmission of knowledge between Antiquity, the Middle Ages and the Renaissance. The text by Alberti, who is both a late Medieval and an early Renaissance man, may provide some clues for the scholars interested in studying the relationship between science and art in the fifteenth century. Alberti does not claim to have invented any kind of measuring system or any new measuring device: he is only transmitting the "playful" – and utmost convenient – mathematical tricks that are useful for solving many problems related to measurements: distances, heights, widths, depths, weights, time, speed, etc., some of which were already present in the sketchbook by the French Villard de Honnecourt some two hundred years before. Alberti is a contemporary of the German Mathes Roriczer whose booklet of 1486 [1977] is well known to scholars, and the comparison between the two texts dating back to the same years points out the importance of Alberti's *Ludi*.

The second text, *Elements of Painting*, is not really a literary text, but rather an outline, consisting in lists of drawing exercises that painters must be able to solve, but Alberti does not show solutions nor explains the procedures to solve them. The text is interesting, however, since it testifies to the mathematical/architectural research in Alberti's day, which was dominated by the progresses in perspective drawing. Stephen

Wassell's commentary focuses on the relationship between three texts: *Elements of Painting, On Painting* (Alberti's better known book) and Euclid's *Elements*. It would indeed be interesting to compare this text also to Piero della Francesca's *De prospectiva pingendi* and to inquire if any exercise is visibly solved in any of the three famous representations of ideal cities that are kept respectively in Urbino, Berlin and Baltimore.

The third text, *On Writing in Ciphers*, is more delightful to read. In this text Alberti unveils part of his professional experience as an Abbreviator at the Vatican, letting us envision the problems of written communication and secrecy in the diplomatic circles. Alberti was asked by a colleague and friend to figure out a formula for coded writing, and so he did. In his commentary to this text Lionel March states that Alberti made use of his skills in coded language while designing the façade of Santa Maria Novella in Florence. Especially, through the use of precise numbers, March says that Alberti was able to hide his signature in the two scrolls screening the aisles roofs. Is this interpretation going to raise some controversy? Where is the limit between science and esoterism? When speaking of medieval or late medieval art and architecture, this limit is always hard to draw. If we agree with March's interpretation, then we must understand the façade as the design of a late medieval philosopher more than that of a Renaissance humanist.

On the contrary, the last text, the very short *On squaring the lune* bears witness to the rebirth of theoretical scientific research. The re-reading of both Euclid's *Elements* and the works by the major ancient mathematicians inspired the Renaissance scientists, who tried to give personal solutions to the three great problems of classical mathematics. Here Alberti is recalling the studies by Hippocrates of Chios on the squaring of the lune. His enthusiasm leads him to conclude that even the circle can be squared. This is a mistake that Alberti and Leonardo da Vinci have in common, as we can see in the sketches collected in the *Codex Atlanticus* (fol. 455r.).

By putting together these different texts, the volume by Williams March and Wassell presents several peculiar aspects of Alberti's character. Being an innovative thematic collection, this series of writings is in itself interesting and useful to many scholars. Further, it offers an original, pioneering translation, which alone makes the book unique. And, on top of that, the commentaries help to appreciate fully both the content and the significance of each text, in relation to the historical period in which it was written.

References

ALBERTI, Leon Battista. 1988. *On the Art of Building in Ten Books*. Joseph Rykwert, Neil Leach and Robert Tavernor, trans. Cambridge MA: The MIT Press.
BELLI, Silvio. 2003. *On ratio and proportion*, Stephen Wassell and Kim Williams, translation and commentary. Fucecchio (Florence): Kim Williams Books (1st ed., 2002).
MARCH, Lionel. 1998. *Architectonics of Humanism: Essays on Number in Architecture*. London: Academy Editions.
RORICZER Mathes. 1977. *Booklet Concerning Pinnacle Correctitude (Büchlein von der Fialen Gerechtigkeit,* 1486). Lon Shelby, trans. In *Gothic Design Techniques: The Fifteenth-Century Design Booklets by Mathes Roriczer and Hanns Schmuttermayer*. Carbondale and Edwardsville: Southern Illinois University Press.
TAVERNOR, Robert. 1998. *On Alberti and the Art of Building*. New Haven: Yale University Press.

About the reviewer

Sylvie Duvernoy is Book Review Editor of the *Nexus Network Journal*.

NEXUS NETWORK JOURNAL Architecture and Mathematics

Subscription information

ISSN print edition 1590-5896
ISSN electronic edition 1522-4600

Subscription rates

For information on subscription rates please contact:
Springer Customer Service Center GmbH
The Americas (North, South, Central America and the Caribbean)
journals-ny@springer.com
Outside the Americas: subscriptions@springer.com

Orders and inquiries

The Americas (North, South, Central America and the Caribbean)
Springer Journal Fulfillment
P.O. Box 2485, Secaucus
NJ 07096-2485, USA
Tel.: 800-SPRINGER (777-4643)
Tel.: +1-201-348-4033 (outside US and Canada)
Fax: +1-201-348-4505
e-mail: journals-ny@springer.com

Outside the Americas

via a bookseller or
Springer Customer Service Center GmbH
Haberstrasse 7, 69126 Heidelberg, Germany
Tel.: +49-6221-345-4304
Fax: +49-6221-345-4229
e-mail: subscriptions@springer.com
Business hours: Monday to Friday
8 a.m. to 8 p.m. local time and on German public holidays

Cancellations must be received by September 30 to take effect at the end of the same year.

Changes of address: Allow six weeks for all changes to become effective. All communications should include both old and new addresses (with postal codes) and should be accompanied by a mailing label from a recent issue.

According to § 4 Sect. 3 of the German Postal Services Data Protection Regulations, if a subscriber's address changes, the German Post Office can inform the publisher of the new address even if the subscriber has not submitted a formal application for mail to be forwarded. Subscribers not in agreement with this procedure may send a written complaint to Customer Service Journals, within 14 days of publication of this issue.

Back volumes: Prices are available on request.

Microform editions are available from: ProQuest. Further information available at: http://www.il.proquest.com/umi/

Electronic edition

An electronic edition of this journal is available at springerlink.com

Advertising

Ms Raina Chandler
Springer, Tiergartenstraße 17
69121 Heidelberg, Germany
Tel.: +49-62 21-4 87 8443
Fax: +49-62 21-4 87 68443
springer.com/wikom
e-mail: raina.chandler@springer.com

Instructions for authors

Instructions for authors can now be found on the journal's website: birkhauser-science.com/NNJ

Production

Springer, Petra Meyer-vom Hagen
Journal Production, Postfach 105280,
69042 Heidelberg, Germany
Fax: +49-6221-487 68239
e-mail: petra.meyervomhagen@springer.com
Typesetter: Scientific Publishing Services (Pvt.) Limited, Chennai, India
Springer is a part of
Springer Science+Business Media
springer.com
Ownership and Copyright
© Kim Williams Books 2011